Synthetic Biology and Greenhouse Gases

Cold Spring Harbor Laboratory Press

SUBJECT COLLECTIONS FROM *COLD SPRING HARBOR*
PERSPECTIVES IN BIOLOGY

Wound Healing: From Bench to Bedside
The Endoplasmic Reticulum, Second Edition
Sex Differences in Brain and Behavior
Regeneration
The Nucleus, Second Edition
Auxin Signaling: From Synthesis to Systems Biology, Second Edition
Stem Cells: From Biological Principles to Regenerative Medicine
Heart Development and Disease
Cell Survival and Cell Death, Second Edition
Calcium Signaling, Second Edition
Engineering Plants for Agriculture
Protein Homeostasis, Second Edition
Translation Mechanisms and Control
Cytokines
Circadian Rhythms
Immune Memory and Vaccines: Great Debates
Cell–Cell Junctions, Second Edition
Prion Biology

SUBJECT COLLECTIONS FROM *COLD SPRING HARBOR*
PERSPECTIVES IN MEDICINE

Combining Human Genetics and Causal Inference to Understand Human Disease and Development
Lung Cancer: Disease Biology and Its Potential for Clinical Translation
Influenza: The Cutting Edge
Leukemia and Lymphoma: Molecular and Therapeutic Insights
Addiction, Second Edition
Hepatitis C Virus: The Story of a Scientific and Therapeutic Revolution
The PTEN Family
Metastasis: Mechanism to Therapy
Genetic Counseling: Clinical Practice and Ethical Considerations
Bioelectronic Medicine
Function and Dysfunction of the Cochlea: From Mechanisms to Potential Therapies
Next-Generation Sequencing in Medicine
Prostate Cancer
RAS and Cancer in the 21st Century
Enteric Hepatitis Viruses
Bone: A Regulator of Physiology
Multiple Sclerosis
Cancer Evolution

Synthetic Biology and Greenhouse Gases

Cold Spring Harbor Laboratory Press

EDITED BY

Daniel Drell
U.S. Department of Energy,
Retired

L. Val Giddings
Information Technology and
Innovation Foundation

Aristides Patrinos
Novim Group, University of
California, Santa Barbara

Richard J. Roberts
New England Biolabs, Inc.

Charles DeLisi
Boston University

Funded by the Bill & Melinda Gates Foundation

COLD SPRING HARBOR LABORATORY PRESS
Cold Spring Harbor, New York • www.cshlpress.org

Synthetic Biology and Greenhouse Gases
Articles online at www.cshperspectives.org

Executive Editor	Richard Sever
Project Supervisor	Barbara Acosta
Editorial Assistant	Danett Gil
Permissions Administrator	Carol Brown
Production Editor	Diane Schubach
Production Manager/Cover Designer	Denise Weiss
Publisher	John Inglis

Front cover artwork: A stylized DNA double helix entwined in a tree trunk emerging from a field of rice ready for harvest illustrates the power of genetics to solve climate challenges. (Image created by Naomi Giddings, NaomiGiddings.com.)

Library of Congress Cataloging-in-Publication Data

Names: Drell, Daniel, editor. | Giddings, Luther Val, editor. | Patrinos, Aristides, editor. | Roberts, Richard J., 1943- editor. | DeLisi, Charles, 1941- editor. | Bill & Melinda Gates Foundation.
Title: Synthetic biology and greenhouse gases / edited by Daniel Drell, U.S. Department of Energy, retired; L. Val Giddings, Information Technology and Innovation Foundation; Aristides Patrinos, Novim Group, University of California, Santa Barbara; Richard J. Roberts, New England Biolabs, Inc.; and Charles DeLisi, Boston University.
Description: Cold Spring Harbor, New York : Cold Spring Harbor Laboratory Press, [2024] | Includes bibliographical references and index. | Summary: "This volume summarizes the proceedings of two workshops sponsored by the Bill and Melinda Gates Foundation on cutting edge systems and synthetic biology research which could facilitate mitigation and adaptation through quantum leaps in photosynthesis, fertilization, methane control, and drought and disease resistance, and thereby substantially increase crop yield and soil carbon. The technologies are at various stages of development--some are ready for field testing while others are still in the laboratory, but those that pass proof of principle should be available and implementable within a decade. A corollary benefit might also be the activation of economic incentives to regard CO_2 and CH_4 as potential resources rather than wastes; this could lower the volume of unproductive arguments over "whose fault it is" that greenhouse gases have increased rapidly in recent times"-- Provided by publisher.
Identifiers: LCCN 2023038119 (print) | LCCN 2023038120 (ebook) | ISBN 9781621825197 (hardcover) | ISBN 9781621825203 (epub)
Subjects: LCSH: Synthetic biology--Congresses. | Greenhouse gas mitigation--Congresses.
Classification: LCC TP248.27.S95 S96 2024 (print) | LCC TP248.27.S95 (ebook) | DDC 363.738/746--dc23/eng/20231003
LC record available at https://lccn.loc.gov/2023038119
LC ebook record available at https://lccn.loc.gov/2023038120

Contents

Foreword

Nothing in biology makes sense except in the light of evolution.
—T. Dobzhansky

SINCE OUR PLANET REACHED CONDITIONS SUITABLE FOR LIFE, nothing on Earth makes sense except in the context of evolution. A combination of physical environments and biological innovations have continuously influenced each other, transforming the geochemical cycles of the planet and life itself. A critical change in the evolution of the web of life was the Great Oxygenation Event. Cyanobacterial photosynthesis provoked mass extinction of the dominant anaerobic microorganisms, priming complex life and triggering a new terrestrial biogeochemical cycle. The ecological network is continuously threatened. Dramatic extinctions have occurred several times.

Another evolutionary branch relevant to our present time is the appearance of Hominini some million years ago, from which we evolved to become the dominant species on Earth. We excel in all the abilities of our ancestors: self-reflection, tool-making, conscious use of plants and other animals, and developing cooperative group actions on a large scale. By the beginning of the nineteenth century, our species' growth rate had increased exponentially. We *Homo sapiens* are such numerous and efficient consumers that we have seized most of the primary products of the planet, causing habitat destruction on a worldwide scale.

Human actions have affected the biogeochemical cycles of the planet, resulting in global warming with severe collateral effects, including ecological catastrophes. Recent natural disasters that have affected the whole world are warning us that we need to act now, using our scientific and social knowledge. Otherwise, such natural disasters will persist and escalate. Plant, microbial, and environmental sciences should be at the forefront of our attention when it comes to technological advancements.

I am confident that human endeavor can find a solution. We can rapidly curb global warming to prevent near-term tipping points while working on long-term strategies. Moving swiftly toward clean energy while changing lifestyle and investing in active greenhouse gases (GHG)-removal technologies are two directions that need to be explored simultaneously. I will first focus on speeding up negative emissions because, even if we move quickly to a low-carbon society, the current level of atmospheric carbon exceeds the capacity of absorption by the Earth's biogeochemical cycles. I will wrap up by highlighting the principles, practices, and behavioral adjustments that mankind as a whole needs to keep our ecosystem healthy.

Without considering plants, it is impossible to think about significant near-term emission reduction and negative emissions technology (NET). Agriculture and other plant-related land uses are a significant source of greenhouse gas emissions, but they are also a source of creative ways to sequester carbon. A sizable portion of the low-cost mitigation for the next 10 years may come from various land-based mitigation strategies. Which strategies are the most appropriate and cost-effective is a topic of discussion. Reducing the conversion and deterioration of natural ecosystems is, in my opinion, the easiest action to take. The protection of forests must be a major concern on a worldwide scale. High-biodiversity forests offer enormous ecological benefits in addition to having a huge storage capacity per unit area. Evaluations of the potential of different land-use strategies based on reforestation or afforestation must work toward not only the long-term security of carbon storage, but also the economic returns and food security. The options will probably be in line with regional requirements and priorities, as well as with countries' capacity to act effectively without jeopardizing other important policy objectives.

I am certain of one thing: we cannot manage land use with the plants we have right now, whether it be for agriculture, afforestation/reforestation, or other purposes. Agricultural and forestry plants have improved agronomic features primarily aimed at high production yields, whereas native plants were designed to survive and thrive in their habitat. Although many commercial trees and crops have genetic variation for these characteristics, selection for low or negative GHG emissions was not on the agenda. Yet, to quickly transfer these features into forestry and agriculture and prevent subsequent ecological catastrophes, plant biotechnology is the only viable option.

Fortunately, more people are finally coming to understand the importance of genetic and genomic technologies to advance nature-based mitigation and adaptation of climate change. Plant biotechnology is ready to offer a variety of tools that will enable a big leap in the rate of discoveries and advances in new traits to reduce global warming and lessen its effects while providing food, feed, fiber, and fuel chains.

The field of plant science is investigating several options to reach zero emissions of greenhouse gases. The most efficient method of storing carbon in terms of time, cost, and scalability is probably carbon sequestration in biomass or soil. Interesting NET techniques include the use of genetic engineering to quickly domesticate perennial crops that have large root systems and to redesign annual crops that have increased root biomass, which reach deeper soil layers and are resistant to degradation.

There are many other viable options, and the best way to discover them is to take the advice of genuine experts. I invite you to read this book with an open mind. It is the outcome of two workshops that brought together a select group of scientists from around the globe to talk about NET-based plants and microbes. This volume serves as proof that we are prepared on a technological level. If the proposed strategies are deployed globally over the next few decades, they could pave the way toward our climate change mitigation goals with co-benefits for human health and a sustainable economy.

There are still gaps in knowledge to be filled. It is in the nature of scientific research that what is known uncovers a vast array of unknowns. Regrettably, the unfavorable perception of plant biotechnology by many people has hindered the advancement of fundamental plant science that is essential today. Specifically, there is only limited knowledge on the pathways utilized by plants to interact with the environment. The physical environment has a significant impact on how plants and microorganisms grow and function, as well as how they interact with each other. The influence of plant–microbe interactions on tolerance to biotic stress, adaptation to drought and temperature, carbon sequestration, and the production of bioactive chemicals and biomaterials has been greatly underestimated. Future crop improvement will require optimized phyllosphere, ectomycorrhizal, and microbiological conditions. For this, interdisciplinary collaboration is required.

Although the obstacles are formidable, there are also good reasons to be hopeful. We have resources, entrepreneurs, and inventors as well as technology. The Bill and Melinda Gates Foundation has taken the lead in promoting agrigenomic initiatives to achieve net-zero greenhouse gas emissions.

How knowledge and the benefits it brings are perceived by the community is as important as the knowledge itself. Technological advances always trigger reactions from some people, and it is important to pay attention to how these reactions can influence the acceptance of an innovative product. But there are also good reasons to be positive about this. I think many people have become aware of the value of biotechnology-based solutions because the effects of climate change are so dire.

I see the Covid-19 crisis as an illustration. We responded to a genuine threat to humanity. Without bypassing any quality or safety related controls, the vaccines were produced in record time. I hope that decision-makers will act appropriately to lessen the regulatory barriers associated with agrigenomics to tackle global warming.

On a less positive note, in the absence of substantial socioeconomic reforms, technology solutions are destined to fall short. To change the game, we must address issues that are currently barely on the public policy radar, such as overpopulation, excessive consumption, and socioeconomic inequality. These are mutually reinforcing threats that greatly contribute to socioeconomic disparity, environmental degradation, and instability. These challenges must be addressed simultaneously, and scientists also have something to say about this.

MARC VAN MONTAGU[1]
Institute of Plant Biotechnology Outreach

---

[1]*Correspondence:* Marc.Vanmontagu@UGENT.BE

Preface

FOR 800,000 YEARS, ATMOSPHERIC CO₂ CYCLED between roughly 130 and 275 ppm. Then came the industrial revolution, and in only 150 years, an instant of geological time, CO_2 spiked to 418 ppm, higher than at any time in the last 3 million years. Moreover, isotopic analysis indicates that the increase is due predominantly to fossil fuels. Similarly, methane and nitrous oxide have spiked to levels that hadn't been seen in millions of years.

Considering these facts, that are just the tip of the iceberg, one can't help being struck by the results of a 2021 Gallup poll indicating that only 57% of U.S. adults believe global warming has an anthropogenic origin. And more than 40% believe its importance is greatly exaggerated, in spite of the increasing number of extreme weather events and other phenomena that have strong anthropogenic footprints. With statistics like those, it's not surprising that large numbers of congressional representatives find legislative proposals easy to ignore or oppose for partisan financial interests.

The good news is that economically viable technologies, which can control and exploit greenhouse gas (GHG) emissions, or even pull GHGs from the atmosphere, are on the horizon. We believe both prevention and pulling will be important if the global temperature is to be kept low enough to avoid serious climatic damage, but current R&D is badly out of balance, with a strong weight on prevention. Prevention is certainly needed, but even if completely successful we would still face a substantial atmospheric surplus of GHGs. In addition, almost all the efforts to draw down atmospheric carbon are focused on traditional engineering and traditional land management. In fact, until 4 years ago, when we held our first conference on the subject, there wasn't much thought, let alone research, on the possibility of applying the emerging and powerful tools of systems and synthetic biology (SSB) to agriculture. Since then, a lot has begun to happen.

This volume follows up two workshops sponsored by the Bill and Melinda Gates Foundation on cutting edge SSB research, which could facilitate mitigation and adaptation through quantum leaps in photosynthesis, fertilization, methane control, and drought and disease resistance, and thereby substantially increase crop yield and soil carbon. The technologies are at various stages of development—some are ready for field testing while others are still in the laboratory, but those that pass proof-of-principle should be available and implementable within a decade. A corollary benefit might also be the activation of economic incentives to regard CO_2 and CH_4 as potential resources rather than wastes; this could lower the volume of unproductive arguments over "whose fault it is" that GHGs have increased rapidly in recent times.

Societal acceptance of new technologies, including often inappropriately applied regulatory measures, is of course another matter, and may well be rate limiting and more difficult to overcome than technical and scientific obstacles. It is important, however, to keep in mind that the known and convincingly terminal consequences of doing nothing may be substantially worse than the risks of thoughtful deployment of new technologies. Although it is impossible to predict what the future will bring, we expect the increasing frequency of severe climate-related events, including heat waves, drought, category 5 hurricanes, and the movement of infectious disease agents to urban environments, will become serious enough threats to health and finance to drive the actions necessary to implement the revolutionary technologies that are sure to develop in the coming decade. We can certainly guide the world toward a better future, but we—and especially the West, where the crisis started—must act now!

DANIEL DRELL, L. VAL GIDDINGS, ARISTIDES PATRINOS
RICHARD J. ROBERTS, CHARLES DELISI

Introduction: Innovations like These Will Help Solve the Climate Crisis

L. Val Giddings

Senior Fellow, Information Technology & Innovation Foundation, Washington, D.C. 20001, USA
Correspondence: vgiddings@itif.org

Hardly a day goes by without another piece praising the potential for gene editing to help solve climate change.[1] Nevertheless, the possible contributions of biology and biotechnology have been conspicuously underplayed in most of the Intergovernmental Panel on Climate Change analyses (IPCC Sixth Assessment Report, www.ipcc.ch/assessment-report/ar6). This is, perhaps, at least partly because it is no small thing to carry an idea from inception to delivery, to navigate the valley of death (between discovery and commercialization), and deliver a functioning solution to a specific problem (Skillicorn 2022, www.ideatovalue.com/inno/nickskillicorn/2021/05/the-innovation-valley-of-death). But this may be starting to change.

In February 2020, an eclectic group of scholars published an important paper examining the range of possible contributions from synthetic biology to climate change solutions (DeLisi et al. 2020). Shortly afterward, the world's leading science policy think tank, the Information Technology and Innovation Foundation (itif.org/publications/2020/01/29/itif-ranked-worlds-top-think-tank-science-and-technology-policy-third), published a report discussing possible gene-edited solutions to climate change challenges (Giddings et al. 2020). Together, these papers paint a hopeful landscape rich with opportunities for solutions to many of the challenges of climate change. Taken together, they beg the obvious question: Which should be pursued first, and how can they best be galvanized? A subset of the authors of these two papers coalesced around this challenge. This volume is one result.

The challenges are not trivial. Since the dawn of the industrial revolution, global atmospheric CO_2 levels have risen from ~280 to 420 ppm (Global Carbon Budget 2022, www.globalcarbonproject.org/carbonbudget/22/files/GCP_Carbon Budget_2022.pdf; The NOAA Annual Greenhouse Gas Index, AGGI, gml.noaa.gov/aggi/aggi.html; Data on CO_2 and greenhouse gas emissions by Our World in Data, github.com/owid/co2-data). This translates to an excess mass of atmospheric CO_2e of approximately 1000 Gt (DeLisi, this volume). So the challenge is twofold—first to reduce emissions to zero; and second, to apply means of rapidly reducing the atmospheric excess below the levels causing major, deleterious climate disruptions. Given the myriad of different anthropogenic and natural sources of greenhouse gases, and the manifold complexities of the global carbon cycle, the challenge is not one problem but many.

This volume follows a series of three international meetings made possible through support from the Bill and Melinda Gates Foundation and the Alfred P. Sloan Foundation, focused on methods for drawing down atmospheric carbon. The first surveyed available methods and included substantial discussion of ethical and regulatory issues (DeLisi et al. 2020). The next two were held in November 2021 at Boston University and February 7–8, 2022, at the U.S. National Academy of Sciences (NAS). Contrib-

---

[1] Portions of this article are reprinted with permission from a piece previously published by the author on the ITIF website: itif.org/publications/2022/08/22/how-agrigenomics-can-help-address-climate-change.

utors to this volume are a subset of the attendees at the last two meetings. The intent of the meetings was to discuss how best to demonstrate proof-of-concept for selected applications and start to flesh out plans to reduce them to practice at scale.

Each of the concepts described has considerable potential, within the foreseeable future, to reduce the amount of greenhouse gases entering the atmosphere, or to reduce the excess carbon now there. Cutting emissions is important but reducing atmospheric carbon concentrations is essential.

Even if humanity succeeds in eliminating carbon emissions altogether—difficult to do while keeping the population fed, housed, and clothed—there is still a large amount of carbon in the atmosphere beyond the capacity of earth's natural ecosystem processes to absorb.

To remove such large amounts of CO_2 from the atmosphere, we need technologies that will deliver at scale—very large scale—and economically. With applications like those described herein, and many more, this is tractable. But to do so, we need to do more than just innovate technologies. As emphasized by Marc Van Montagu in the Foreword (and in the chapter in this volume), we need to remove obstacles to such innovations in policies, law, and regulation. It's time for countries around the world, including the United States, and especially the member states of the European Union, to set aside laws and regulations left over from the early days of the biotechnology revolution. They were driven by fears that have long been shown unfounded and a lack of understanding now thoroughly remedied by experience and must be replaced by policies that will enhance and enable innovations rather than impede them. The world waits.

The contributions in this volume describe specific projects with the potential to sequester or reduce emissions of greenhouse gases considerably and estimate the costs involved and the impact were the project deployed at scale. It is important to note that there are multiple opportunities for the projects to synergize with each other and other initiatives.

Charles DeLisi analyzes the expected impact of the $390 billion climate change component of the U.S. Congress' Inflation Reduction Act H.R. 5376 (www.democrats.senate.gov/imo/media/doc/inflation_reduction_act_of_2022.pdf). Straightforward projections indicate that even if scaled to worldwide participation, the impact on atmospheric CO_2 by the end of the decade will be inconsequential. This result highlights the need to include CO_2 drawdown in any climate-mitigation/adaptation strategy, and to provide substantially greater support for creative solutions to energy generation.

Julie Gray describes research in her laboratory, which has successfully produced rice strains with fewer stomata and thus a decreased need for water. This is important for both adaptation and mitigation. With respect to the latter, water-intensive cultivation of rice leads to degradation of soil organic matter by methanogens and other fermenting bacteria, a process that releases methane and accounts for some 2.5% of anthropogenic CO_2 equivalent emissions (Mboyerwa et al. 2022).

These emissions could be substantially alleviated by optimizing crops to reduce water loss and speed the much-needed shift from paddy rice toward dry seeded systems (Harlan and Pitrelli 2022; Caine et al. 2023). In collaboration with the International Rice Research Institute in the Philippines, novel rice varieties with reduced stomatal density have already been produced through gene editing and trait selection that could be field tested rapidly and adopted by farmers in the near future. The combined impact of enhanced rice drought tolerance, together with reduced methane emissions and energy savings from irrigation could reduce CO_2 emissions by more than a Gt per year, with further reductions possible by application to other major food crops including maize, soybean, and rice as well as biomass crops like poplar and switchgrass.

Mary Lidstrom describes a plan for methane capture from air using aerobic methanotrophs. Methane has a warming impact 34 times greater than CO_2 (on a 100-year timescale; 86 times greater on a 20-year timescale) and a relatively short half life in the atmosphere (~10 years). This makes it a key short-term target for slowing global warming by 2050. Lidstrom's team is developing biofilter technology using bacteria that metabolize methane to remove methane from the air over emission sites, including agricultural areas, where methane is enriched compared to the atmosphere as a whole. This technology will be designed also to reduce production of another greenhouse gas, N_2O, by limiting nitrate availability. By targeting tens of thousands of such sites, enhancing the current capacity of such technology, and envisioning a 20-year deployment, it would be possible to remove a total of 0.3 Gt methane (10 Gt CO_2e) by 2050 (0.5 Gt CO_2e/yr), resulting in 0.2°C less global warming.

Lisa Stein is working with biological nitrification inhibitors and soil-free systems. While the Haber–Bosch process for N-fixation has enabled a stable food supply for half of humanity, the heavy use of synthetic fertilizers has caused a radical imbalance in the global N-cycle (Smil 1999). The resulting increases in nitrate production and GHG emissions have contributed to eutrophication of ground and surface waters, growth of oxygen-depleted zones in coastal regions, ozone depletion, and has exacerbated rising global temperatures. According to the Food and Agriculture Organization of the United Nations, agriculture releases approximately 9.3 Gt CO_2 equivalents per year, of which methane and nitrous oxide account for 5.3 Gt CO_2e, ~10% of the total. N-pollution and slowing the runaway N-cycle requires a combined effort to replace chemical fertilizers with nitrification inhibitors based on biological formulations, which could significantly reduce agriculture-related GHG emissions, protect waterways from nitrate pollution, and soils from further deterioration. Stein's team aims to bring biological nitrification inhibitors (BNIs) to the market to curb the microbial conversion of fertilizer nitrogen into greenhouse gases and other toxic intermediates. Worldwide adoption of these plant-derived BNI molecules in combination with biological fertilizers would substantially elevate nitrogen use efficiency (NUE) by crops while blocking the dominant source of nitrous oxide to the atmosphere. In addition to biological fertilizers and BNIs, a second project to curtail N-pollution, soil erosion, and deterioration of freshwater supplies pursues the development of improved microbial inocula to increase NUE without GHG production in soil-free hydro- and aquaponics operations. The carbon cycle in these systems can be closed by including anaerobic digestion of solid waste followed by microbial conversion of methane into single-cell protein for fish feed.

Ken Nealson describes a project involving microbial acceleration of the natural process of silicate-carbonate mineral weathering as a method of atmospheric CO_2 drawdown. There is general agreement that rock weathering via the conversion of basaltic (silicate) rocks to limestone-like carbonates offers an excellent (and irreversible) way to reduce atmospheric CO_2 levels. The natural rates of such weathering reactions are slow, but Project Vesta is developing living microbial catalysts that greatly enhance the rate of this weathering. Preliminary work has shown major rate enhancement(s), and the project is now ready to begin screening and genetically engineering microbes to maximize the rate(s) at which they accelerate silicate weathering. The next step will be to scale up to 3000-liter batches, followed by medium and then large-scale reactors. Vesta anticipates about 5 million tons of carbon removal per year with current supply partners, and after that there are sufficient mineral reserves and mine tailings to scale beyond the gigaton level in less than a decade. Their approach will encompass coastal, terrestrial (soil), and freshwater mineral enhancement, and include both reactor-based and in situ systems. Project Vesta has the capability to bring the described approach(es) to scale rapidly, using several laboratories and field stations where research is ongoing, together with an outstanding scientific, engineering, and administrative staff, already working on permitting and other business issues.

Tobias Erb leads a large team looking for ways to improve photosynthesis beyond its natural limits. Plants generally use only about 1% of the sunlight that falls on them to make carbohydrates, consuming or "fixing" atmospheric carbon in the process. Research teams around the world are exploring ways to improve natural photosynthesis and have made considerable progress to date. But Erb and his colleagues are poised to make a quantum leap by radically reinventing photosynthesis using synthetic biology, enzyme engineering and machine learning to create innovative crops featuring a new-to-nature CO_2-fixation metabolism with photosynthetic yields increased by 20% to 60% (potentially up to 200%). They have already demonstrated several synthetic pathways for improved CO_2 fixation that are up to 20× faster than natural photosynthesis and require 20% less energy in the laboratory. Erb's team is working to implement these new-to-nature solutions in microorganisms and plants to improve their CO_2 uptake efficiency. Models suggest that their approach could lead to the ability to sequester upward of 3 Gt of CO_2e annually, if applied to crops alone.

Xiaohan Yang proposes a three-pronged mutually reinforcing approach to carbon sequestration and climate change adaptation. This would involve increased photosynthesis, increased translocation of captured carbon, and increased soil capacity—and would, if successful, have a major impact.

The three objectives are (1) integrative engineering of CO_2 capture, storage, and utilization in fast-growing poplar (which can be used as a feedstock for biofuels, biomaterials, and engineered for deeper roots and root architectures for increased carbon storage), (2) genetically enhanced agave-mediated carbon sequestration and utilization in dry and hot regions, and (3) the development of "care-free/climate-friendly" lawn grasses. It is noteworthy that increased production of agave would not only contribute to amelioration of malnutrition, which will become increasingly serious as the climate changes, but it would enable utilization of the planet's 500 million acres of marginal land (arid and semi-arid) for production of agave and poplar, and for the sequestration of 18 Gt CO_2e per year of atmospheric carbon. In addition, the crassulacean acid metabolism (CAM) photosynthesis pathway, which would be utilized, has the unique feature of high water-use efficiency due to daytime closure of stomata for reducing water loss mediated by transpiration, and night-time opening of stomata for CO_2 uptake. More gen-

erally, CAM genes could also be engineered into C3 plant species to increase drought tolerance.

Applying "care-free/climate-friendly" varieties of perennial ryegrass, tall fescue, Kentucky bluegrass, and fine fescue that are used in lawns, athletic fields, golf courses, etc., will lead to significant reductions in mowing frequency, fertilizer inputs, and water use, and could result in a greater than 90% reduction in emission of CO_2, nitrous oxide, and methane associated with lawn care. Several studies have shown that greenhouse gas emission from lawn care, which includes fertilizer and pesticide production, watering, mowing, and other lawn care practices, is much greater than the amount of carbon stored by lawn grasses. Yang estimates that in the United States alone, lawn-care practices contribute at least of 41.3 million metric tons of CO_2e.

Forest Biotech company FuturaGene (parent company Suzano SA, Brazil) is focused on the sustainable enhancement of renewable plantation forest species. Plantation forests constitute only 7% of global forest area, but they provide ~50% of industrial wood. Shortfalls of 1–4 billion m^3 in industrial roundwood supply are projected by 2050, and the productivity of planted forests must triple by 2050 if global climate mitigation and adaptation targets are to be supported and destruction of natural forests prevented. FuturaGene proposes within the next 10 years to deploy genetic modification (GM) technologies (including direct yield enhancement and photorespiratory bypass) to plantation forestry across the subtropics to address those targets and they have the ability through parent company Suzano SA to do so at scale. FuturaGene has developed and obtained regulatory approval (in Brazil) for a direct yield-enhanced eucalyptus variety, which is ready for landscape-level testing (Riquelme 2015). Current carbon sequestration is estimated to be around 240 tons per hectare during a 7-year growth cycle (a 2015 desk study). A 12% increase in yield by either route would deliver 270 tons/ha/7 years. If yield enhanced lines were planted over the entire 1.2 million ha estate of Suzano, this would correspond to 324 million tons of CO_2e per cycle. Across the entire Brazilian estate of 9 million hectares, this would be 2.4 GtCO_2e per cycle.

In my own contribution, I point out that while projects like these hold considerable GHG mitigation potential, the rate-limiting factor in their ability to sequester carbon is implementation, which is contingent on public acceptance. Policies, regulations, and business practices such as industry certification all create considerable barriers to development and deployment of innovative solutions developed with gene editing and genetic engineering. This is the case around the world despite the lack of scientific justification for such discrimination and in the face of massive experience demonstrating superior safety and sustainability of such technologies, and the lack of commensurate benefit from impeding policies and regulations. If the acceptance climate is not improved, these solutions will not be deployed. Pushing back against the concerted disinformation campaigns from special interests that have driven such discriminatory policies is difficult, particularly for governments, but independent, science-based voices are uniquely suited for the task. There are several entities with proven track records in this space, a handful of which are listed below.

- The Genetic Literacy Project (GLP), geneticliteracyproject.org;

- The Institute for Food Agricultural Literacy (IFAL), ifal.ucdavis.edu;

- The International Service for the Acquisition of Agri Biotech Applications (ISAAA), www.isaaa.org/kc/default.asp;

- PG Economics, pgeconomics.co.uk/publications; and the

- Information Technology and Innovation Foundation (ITIF), itif.org/issues/agricultural-biotech.

There is agreement across several proposals that the optimal way to reduce CO_2 is to improve biological productivity through crop engineering and increasing carbon fixation rates. There is also recognition that nitrogen use efficiency, as mediated by the soil microbiome, is essential to achieving this goal. To further decrease GHGs outside of crop systems, microbial processes including atmospheric methane oxidation and siderophore-mediated silicate weathering have been proposed. Organization to leverage the outputs of the projects described could center around the plant–microbe–geology axis with feedback between the atmosphere and hydrosphere. The idea would be to integrate implementation and monitoring of the proposed technologies to incorporate and include synergistic (and unintentional) effects across other ecospheres (e.g., biosphere, atmosphere, geosphere, hydrosphere). An organizational strategy that interconnects across systems would allow additional projects to be incorporated so long as the intention is to remain conscious of the interconnections and feedback. This strategy will also force us to keep in mind that systems necessarily work together and affect one another. We cannot afford to continue the age-old practice of changing one component of a system with the belief that nothing else will change in response!

Together, the total carbon sequestration capability of the projects described in this volume is estimated to begin at ~18 Gt CO_2e. Given potential synergies and cooperation between them, supporting them as a group would yield greater benefits than selecting individual projects from a menu of discrete offerings, and sustained support over time with minimal bureaucratic burden would lead to the greatest returns on investment. None appears to be unduly hazardous. To quantify and evaluate their risks with precision is not easily done. It is important to remember, however, that the decisions that must be taken to develop and deploy such measures do not require understanding the absolute risks they may entail. These decisions depend instead on ascertaining the relative risks compared with the alternatives, including inaction. In medicine, high-risk experimental therapies can be justified when the alternative is terminal; we need to consider that we are on a road to irreversible and insufferable planetary warming unless we rapidly undertake bold and aggressive corrective actions.

We are grateful to all the contributors to this volume, which is made freely available electronically by generous financial support from the Bill and Melinda Gates Foundation. We also particularly want to acknowledge the support and encouragement of Rodger Voorhies, president, Global Growth and Opportunity Fund.

REFERENCES

Caine RS, Harrison EL, Sloan J, Flis PM, Fischer S, Khan MS, Nguyen PT, Nguyen LT, Gray JE, Croft H. 2023. The influences of stomatal size and density on rice abiotic stress resilience. *New Phytol* **237**: 2180–2195.

DeLisi C, Patrinos A, MacCracken M, Drell D, Annas G, Arkin A, Church G, Cook-Deegan R, Jacoby H, Lidstrom M, et al. 2020. The role of synthetic biology in atmospheric greenhouse gas reduction: Prospects and challenges. *BioDesign Res* **2020**. doi:10.34133/2020/1016207

Giddings LV, Rozansky R, Hart DM. 2020. Gene editing for the climate: biological solutions for curbing greenhouse emissions. https://www2.itif.org/2020-gene-edited-climate-solutions.pdf

Harlan C, Pitrelli S. 2022. A drought in Italy's risotto heartland is killing the rice. https://www.washingtonpost.com/world/2022/07/22/italy-drought-2022-rice-risotto

Mboyerwa PA, Kibret K, Mtakwa P, Aschalew A. 2022. Greenhouse gas emissions in irrigated paddy rice as influenced by crop management practices and nitrogen fertilization rates in eastern Tanzania. *Front Sustain Food Syst* **6**: 868479. doi:10.3389/fsufs.2022.868479

Riquelme L. 2015. FuturaGene's genetically modified eucalyptus approved in Brazil. https://www.labiotech.eu/trends-news/futuragenes-genetically-modified-eucalyptus-approved-in-brazil

Smil V. 1999. Detonator of the population explosion. *Nature* **400**: 415. doi:10.1038/22672

South PF, Cavanagh AP, Liu HW, Ort DR. 2019. Synthetic glycolate metabolism pathways stimulate crop growth and productivity in the field. *Science* **363**: peaat9007. doi:10.1126/science.aat9077

An Agrigenomics Trifecta: Greenhouse Gas Drawdown, Food Security, and New Drugs

Charles DeLisi

College of Engineering, Boston University, Boston, Massachusetts 02215, USA

Correspondence: Delisi@bu.edu

An abundance of data, including decades of greenhouse gas (GHG) emission rates, atmospheric concentrations, and global average temperatures, is sufficient to allow a strictly empirical evaluation of the U.S. plan for controlling GHGs. This article presents an analysis, based solely on such data, that shows that the difference between atmospheric GHG levels that will be reached if current trends continue, and levels that would be achieved if the goals of the plan are met—even with worldwide implementation—is inconsequential. Further, the expected globally averaged temperature differences are well within measurement error. The results lend additional support to the argument that any mitigation strategy must include drawdown of atmospheric GHGs. Equally important, a particular drawdown strategy, agrigenomics, offers the opportunity for a revolutionary trifecta: climate change mitigation, food security, and medical advances.

Global warming is incontrovertible, and confidence is very high that the extreme events witnessed in recent decades are mainly anthropogenic in origin (IPCC 2021). The confident linkage of a changing climate to human activity, rather than to natural and unknown causes, is crucially important since it means that we can in principle reduce warming by limiting human-generated greenhouse gases (GHGs). Unfortunately, although the world has witnessed three decades of multinational meetings and agreements, the average global temperature continues to increase, as do its many and increasingly disruptive correlates.

One might take heart that in 2022, the United States Senate passed the Inflation Reduction Act (Science for Impacts, Resilience and Adaptation [SIRA]), which included some $391 billion for climate change mitigation and adaptation. It has been described as the most ambitious climate action ever taken by Congress and, sadly, it no doubt is.

EMISSION REDUCTION ALONE WILL NOT CONTROL GLOBAL WARMING

Current Policy

Recent estimates indicate that SIRA might reduce GHG emissions to a level that is 32%–40% below 2005 levels by the end of the decade (Jenkins et al. 2022). Before commenting on the consequences of such a reduction, it is important to note that the warming potential of

GHGs emitted by the United States in 2005 was equivalent to that of ~6.4 billion metric tonnes (Gt) of carbon dioxide (CO_2e), whereas by the end of 2019 emissions had declined to ~5.8 Gt (Ritchie and Roser 2020).

A 40% goal, to be optimistic, would require reducing emissions by 0.36 Gt/yr over a 7-yr period. Since the United States currently accounts for only ~12% of global emissions of CO_2e, much of the potential impact of the legislation, even if it proved effective, would be lost. I will, therefore, assume, with unwarranted optimism, that the rest of the world adopts an SIRA-like policy, and comment on the impact under that condition. The target for the average annual global emission reduction in CO_2e would then be ~3 Gt/yr = 0.43 ppm/yr.[1]

The SIRA-based plan immediately suggests three questions: What will the atmospheric concentration be if it is globally successful; How much of an impact will the reduction in CO_2e have, relative to business as usual (BAU), on the average global temperature; and What will the cost be per ppm of CO_2e reduction? The first question has been addressed, at least in terms of the U.S. contribution, by detailed and careful analysis of different decarbonization scenarios (REPEAT Project, repeatproject.org/policies/net-zero). The philosophy adopted here is not to forecast the reduction, but to ask for the implications of a prespecified reduction, as embodied in these questions. I believe informative approximate answers can be obtained based on reasonable assumptions and on purely empirical data (i.e., no theory).

The results presented here suggest that the impact of SIRA on climate change mitigation is essentially inconsequential and add urgency to introducing the kinds of action needed to mitigate climate change. Specifically, the final section of this article includes a discussion of agrigenomics as an additional and important component of a global policy that could, if well organized and funded at scale,

substantially mitigate a problem that all available data indicate continues to grow worse at an alarming rate. Moreover, agrigenomics offers the possibility of a multiwin strategy, including major benefits to multiple economic sectors.

Analysis

There are several auxiliary quantities that are required to obtain answers: the annual BAU rate of increase in global CO_2e emission, defined as ξ_1; the annual rate of reduction, ξ_2, required by a specified reduction in CO_2e emission; an empirical relation based on historical data between atmospheric CO_2e and average global temperature, and the fraction of CO_2e emission that remains in the atmosphere, also based on historical data.

Useful estimates of ξ_1 and f, the fraction of CO_2e that remains in the atmosphere, can be made if changes in concentration vary linearly with time. As shown in Figure 1, the rate at which atmospheric CO_2e is increasing has satisfied that condition for at least the last four decades. The linear increase implicitly averages over a rate of accumulation minus a rate of loss for the various GHGs. Although the averaging time is small compared to the atmospheric residence time of most GHGs, especially CO_2, the GHG increase is more or less in a steady state for the relevant period.

In contrast to the implicitly averaged atmospheric concentration increase, annual emission prior to 2007 changes erratically and nonlinearly (Fig. 2; Ritchie and Roser 2020). Consequently, emission data prior to 2007 are not useful. For the 13-yr period from 2007 to 2019 inclusive, the slope of CO_2e emission versus year is reasonably linear, and a least-squares fit gives $\xi_1 = 0.085$ ppm/yr. As indicated above, $\xi_2 = -0.43$ ppm/yr.

The fraction, f, of emitted CO_2e that remains in the atmosphere was estimated by dividing the change in atmospheric concentration during the 2007–2019 period (40 ppm) by the sum of the annual emissions during that period (93 ppm) to obtain $f = 0.43$.

---

[1] The conversion 6.9 GtCO_2eq = 1 ppm was obtained by summing the conversion factors for CO_2, CH_4, and N_2O, and F gases, weighted by their concentrations normalized to 1. It, therefore, omits ~4% of GHGs.

Cite this article as *Cold Spring Harb Perspect Biol* doi: 10.1101/cshperspect.a041676

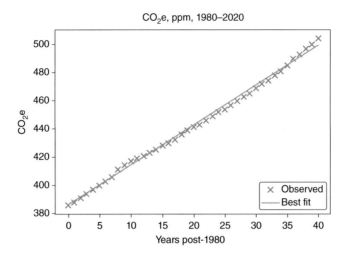

CO$_2$e, ppm, 1980–2020

Figure 1. Carbon dioxide equivalent (CO$_2$e) atmospheric concentration from 1980 to 2019 has increased linearly at a rate of 2.84 ppm/yr. (From the National Oceanic and Atmospheric Administration [NOAA] annual greenhouse gas index, gml.noaa.gov/aggi/aggi.html.)

The Atmospheric Concentration Difference between SIRA and BAU Is Roughly 1%

Let c$_n$ denote the carbon dioxide equivalent concentration n years after year 0 (see Box 1),

$$c_n = c_0 + nf\Delta + \frac{nf(n+1)\xi_j}{2}; \quad j = 1, 2, \quad (1)$$

and δ, the difference between concentrations for SIRA and BAU,

$$\delta = \frac{fn(n+1)(\xi_1 - \xi_2)}{2}, \quad (2)$$

where Δ is the emission rate, and c$_0$ atmospheric concentration, in year 0.

From Equation 2, we find that SIRA, if globally applied, will decrease the atmospheric concentration relative to BAU by 6.2 ppm. To place this in perspective, it is ~1.2% of the 2019 atmospheric CO$_2$e (500 ppm).

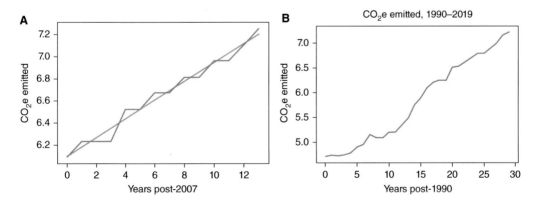

Figure 2. (A) Carbon dioxide equivalent (CO$_2$e) emitted annually, 2007–2019 inclusive. The slope $\xi_1 = 0.085$ ppm/yr. The fraction f, of CO$_2$e emitted that remains in the atmosphere, was estimated by dividing the total emitted CO$_2$e during the period shown by the increase in atmospheric concentration (Fig. 1) during the same period. (B) CO$_2$e emitted annually, 1990–2019 inclusive (Ritchie and Roser 2020).

The importance of CO_2 to the total GHG concentration cannot be overstated. CO_2 not only consists of some 72% of annual GHG emissions, it does not react chemically, has no natural sinks, and therefore has a long and complicated atmospheric residence (Fig. 3; Ritchie and Roser 2020). Under SIRA, CO_2 would reach ~431 ppm, or ~3.4 trillion tonnes (Tt)—that is, an excess of about $(431 - 280) \times 7.8 = 1.2$ $TtCO_2$ above preindustrial levels. Even if emission stopped immediately, the excess atmospheric CO_2 in 500 years would be in the neighborhood of 330 Gt.

Temperature Differences Are Inconsequential

A linear least-squares fit of temperature to concentration (Fig. 4) has a slope of $0.0068 C \cdot CO_2 e^{-1}$ (read 0.0068°C/ppm of $CO_2 e$). This gives a temperature deviation between BAU and SIRA of ~0.04°C, which is within measurement uncertainty. There should be no need to emphasize, but I will nevertheless remark, that if the United States acts alone, ξ_2 would have to be multiplied by 0.12 used for estimating the share of U.S. $CO_2 e$, and the temperature deviation from BAU would be essentially meaningless.

Costs

An estimate of cost does not involve a complex economic analysis: consistency requires that it be obtained by multiplying the SIRA budget using the $CO_2 e$ scaling factor. In particular, $391 billion/0.12 gives 3.26×10^{12} or (dividing by 6.2 ppm) $526 billion/ppm of $CO_2 e$ emission reduction relative to BOA. That translates to ~$76/tonne of $CO_2 e$, which is within the range of estimates of carbon abatement costs. It is important to be clear about what this money is buying—it is providing jobs in an important economic sector, and it is appropriately part of the Economic Recovery Act; it is not, however, doing anything to mitigate global warming.

We can hope that other wealthy nations will do their part in reducing GHGs, but hope is not policy, and for a plan such as SIRA, the impact would be far too small even if the world cooperated. Moreover, although SIRA sounds bold, it is not—and if, as some say, it is a step in the right direction, it is a political step, not a technological step. In fact, to the extent that it gives false hope and thereby reduces pressure for consequential action, it may be worse than doing nothing.

The reasons for the disappointing lack of consequence are not difficult to identify. One is that, unlike the majority of nations for which GHG emissions have increased since 2003, the U.S. GHG emissions have declined. Consequently, using 2003 as a point of reference can be misleading; it would seem more straightforward to speak in terms of a reduction below current levels. In addition, exuberance over an apparitional reduction in emission rate can blur what should be a clear vision of a planetary reality, that is, the continued rapid emission and accumulation of $CO_2 e$.

THE PATH FORWARD MUST INCLUDE ATMOSPHERIC GHG REMOVAL

To be clear, I am not saying that current strategies using alternative energy sources, green industries, economic stimuli, and emission constraints should be abandoned. On the contrary, as argued previously (DeLisi 2019; DeLisi et al. 2020), they must be included as part of an integrated strategy, but one that includes a strong GHG drawdown component. In particular, the results indicate that under a reasonable set of circumstances, neither strategy alone effectively reduces atmospheric GHGs, but together the impact is substantial. They also indicate an unusually strong effect of a decadal delay in the start of an effective strategy (details are in DeLisi 2019).

A Strategy that Includes Agrigenomics Offers the Possibility of a Revolutionary Trifecta

GHG drawdown is an engineering problem and numerous solutions have been proposed, some of which have been under consideration for decades (National Academies of Sciences, Engineering and Medicine 2019). Until recently, however, the approach that we believe has the greatest potential, agrigenomics, has received very little attention. To a first approximation, agrigenomics can be thought of as traditional land management on steroids. But as the contributions to this collection demonstrate, its advantages are more than just quantitative: it en-

 Cite this article as *Cold Spring Harb Perspect Biol* doi: 10.1101/cshperspect.a041676

BOX 1. DERIVATION OF EQUATION 1

The simplest way to obtain Equation 1 is to write out expressions for the concentration in the first few years.

The atmospheric concentration at the end of 2020 for BAU ($j = 1$) and SIRA ($j = 2$) is

$$c_1 = c_0 + f(\Delta + \xi_j) \quad j = 1, 2,$$

where $f(\Delta + \xi_j)$ is the concentration increase in year 2020.

The amount emitted in 2021 is the amount emitted in 2020 + the change in emission in 2021,

$$(\Delta + \xi_j) + \xi_j = \Delta + 2\xi_j.$$

Therefore,

$$c_2 = c_1 + f(\Delta + 2\xi_j) = c_0 + f(\Delta + \xi_j) + f(\Delta + 2\xi_j) = c_0 + 2f\Delta + 3f\xi_j.$$

The emission in 2022 is the emission in 2021 + the change in emission in 2021,

$$\Delta + 2\xi_j + \xi_j = \Delta + 3\xi_j,$$

$$c_3 = c_2 + f(\Delta + 3\xi_j) = c_0 + 2f\Delta + 3f\xi_j + f(\Delta + 3\xi_j) = c_0 + f(3\Delta + 6\xi_j).$$

The relation between c_n and c_0 can now be seen to have the form,

$$c_n = c_0 + nf\Delta + fr_n\xi_j,$$

where $r_n = 1, 3, 6, 10, \ldots$ for $n = 1, 2, 3, 4, \ldots$ r_n will be recognized as the sum of the first n integers:

$$r_n = \sum_{k=1}^{n} k.$$

Therefore, $r_n = n(n + 1)/2$,

$$c_n = c_0 + nf\Delta + \frac{fn(n + 1)\xi_j}{2},$$

and the atmospheric concentration difference, δ, between BAO and SIRA after n years follows immediately:

$$\delta = \frac{fn(n + 1)(\xi_1 - \xi_2)}{2}.$$

ables possibilities that were previously difficult even to imagine. Perhaps most importantly, it offers the possibility of a triple win, including new approaches to food security in response to a changing climate, and the possibility of discovering new generations of promising compounds for pharmaceutical development. And, of course, substantial economic benefits derive from all three.

The biological approach that has received the most attention is bioenergy with carbon capture and sequestration (BECCS). The idea is to plant and then remove trees after a period of rapid growth, followed by burning to provide energy, capturing and sequestering the released carbon, and then continuing the sequence with another round of seedlings. Trees of course require land, and, unfortunately, the land require-

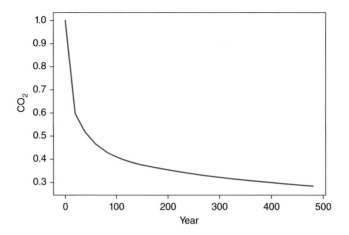

Figure 3. Approximately 28% of a pulse of CO_2 is still in the atmosphere after 500 years. Removal follows a multiexponential decay with three components, whose driving processes are not well identified. (Amplitudes and time constants data are from Shine et al. 2005.)

ment for even a moderate reduction in emission, let alone a reduction in the concentration of atmospheric GHGs, is essentially unattainable. Other biomass energy/carbon removal methods are also being explored (DeLisi et al. 2020), but they all face serious scaling difficulties.

I mention BECCS even though it is not genomically based, because of its visibility and because agrigenomic advances can lower the threshold for its use by substantially improving land use. Efficient land use, of course, has numerous benefits beyond just boosting bioenergy and carbon removal efforts. For instance, agrigenomic advances can enable the cultivation of crops in difficult environments through genetic modification to increase drought and salinity tolerance (Caine et al. 2019). And of course, an increase in photosynthetic efficiency, such as by the creation of photorespiration pathways that are new to nature (Scheffen et al. 2021; Erb et al.

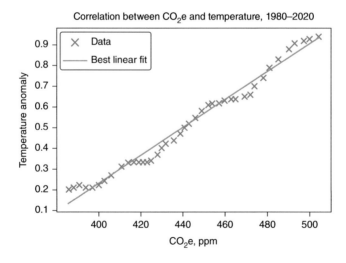

Figure 4. Global average temperature has been increasing approximately linearly with carbon dioxide equivalent (CO_2e) concentration for more than four decades. (Temperature data are from NASA/GISS, climate.nasa.gov/vital-signs/global-temperature.)

 Cite this article as *Cold Spring Harb Perspect Biol* doi: 10.1101/cshperspect.a041676

2023), could have an enormous impact. Futura-Gene has already engineered eucalyptus trees with a 30% increase in yield (May et al. 2022). Here again, the idea would be to grow and replace high-yield trees, and to thereby increase CO_2 removal. Moreover, the possibility of moving beyond eucalyptus to tress whose wood is high on the Janka scale,[2] for example, at or above the hardness of Brazilian ebony, could provide a source of building material that would substantially reduce the carbon footprint of cement production.

Another promising area of research is the use of mitochondrial editing to modify the active site of rubisco in C3 plants, potentially increasing the photosynthetic efficiency and carbon-to-oxygen ratio. Finally, toxicity and other damage that accompanies the use of excess fertilizer can be reduced by increasing the efficiency with which microbes fix atmospheric nitrogen.

The third arm of the trifecta is medical. Until the relatively recent ascendency of structural chemistry, living organisms, and especially plants, were the main source of drugs. Currently, some 11% of the drugs classified as essential by the World Health Organization originated in plants (Veeresham 2012) including digitalis, ergotamine, quinine, and salicylates (Sen and Samanta 2015).

Indeed chemical methods have not so much displaced natural products as the source of drugs, as they have accelerated the opportunities for exploiting plants in the search for new medicines. Thus, sequencing enables rapid, inexpensive elucidation of structure and the structures of enzymes involved in their synthesis, design enables increased potency, and low-cost synthesis enables amplification of availability. An example is the vinca alkaloids that are widely used to treat various cancers (Sears and Boger 2015).

Considering the diversity and magnitude of the plant kingdom, it would be surprising if there were not a treasure trove of untapped opportunities that could be realized with accelerated efforts in agrigenomics. These include the potential for the discovery of enzymes crucial to metabolism, of new pathways for secondary metabolites, and transgenic plants with specified traits of medical benefit.

It should be evident from these remarks and from the other articles in this collection that scientific and technological opportunities for climate modulation abound. There are also, of course, multiple possible roadblocks. Most are too complex to be discussed here, but two in particular must be mentioned if this monograph is to have even the semblance of completeness. Regulatory and other related social issues were discussed in depth at our 2020 conference on Climate Change and Systems Biology and are summarized in DeLisi et al. (2020) and Giddings (2023).

With respect to organization, although there is plenty going on scientifically, a challenge as substantial as climate change is not likely to be met by scattered efforts here and there; it is best met by a well-thought-through cohesive plan. Highly complex challenges encountered in the past—the Manhattan Project, the moonshot, the Human Genome Project—were all met because science and technology were supported by superb large-scale organizational leadership. Climate change, with its mix of politics, economics, ethics, cultural diversity, and technology, is, in my opinion, considerably more complicated than any technological challenge faced by previous generations. It is past the time that some nation, or group of billionaires willing to pool resources, step forward with a plan to kickstart a historic effort. If that happens a parade will follow, if for no other reason than fear of being left behind economically.

ACKNOWLEDGMENTS

This article has been made freely available online by generous financial support from the Bill and Melinda Gates Foundation. We particularly want to acknowledge the support and encouragement of Rodger Voorhies, president, Global Growth and Opportunity Fund.

---

[2]The Janka scale quantifies the force required to drive a standardized steel ball halfway into a piece of wood under standardized conditions. For example, rosewood and hickory require 7900 and 8100 N, respectively, Brazilian ebony 16000 N, and wood with the very highest indices ~22000 N.

REFERENCES

Reference is also in this subject collection.

Caine RS, Yin X, Sloan J, Harrison EL, Mohammed U, Fulton T, Biswal AK, Dionora J, Chater CC, Coe RA, et al. 2019. Rice with reduced stomatal density conserves water and has improved drought tolerance under future climate conditions. *New Phytol* **221:** 371–384. doi:10.1111/nph.15344.11

DeLisi C. 2019. Correction to: The role of synthetic biology in climate change mitigation. *Biol Direct* **14:** 24. doi:10.1186/s13062-019-0254-9

DeLisi C, Patrinos A, MacCracken M, Drell D, Annas G, Arkin A, Church G, Cook-Deegan R, Jacoby H, Lidstrom M, et al. 2020. The role of synthetic biology in atmospheric greenhouse gas reduction: prospects and challenges. *BioDesign Res* **14:** 14.

* Erb TJ. 2023. Photosynthesis 2.0: realizing new-to-nature CO_2-fixation to overcome the limits of natural metabolism. *Cold Spring Harb Perspect Biol* doi:10.1101/cshperspect.a041669

* Giddings LV. 2023. Overcoming obstacles to gene-edited solutions to climate challenges. *Cold Spring Harb Perspect Biol* doi:10.1101/cshperspect.a041677

IPCC. 2021. Summary for policymakers. Climate change 2021: the physical science basis. *In Contribution of Working Group I to the Sixth Assessment Report of the Intergovernmental Panel on Climate Change.* Cambridge University Press, Cambridge.

Jenkins JD, Mayfield EN, Farbes J, Jones R, Patankar N, Xu Q, Schivley G. 2022. *Preliminary report: the Climate and Energy Impacts of the Inflation Reduction Act of 2022.* REPEAT Project, Princeton, NJ. https://repeatproject.org/docs/REPEAT_IRA_Prelminary_Report_2022-0804.pdf.

May M, Hirsch S, Abramson M. 2022. *How can agrigenomics help address climate change?* Rockefeller Philanthropy Advisors Workshop, Washington, DC.

National Academies of Sciences Engineering and Medicine. 2019. *Negative emissions technologies and reliable sequestration: a research agenda.* The National Academies Press, Washington, DC.

Ritchie H, Roser M. 2020. CO_2 and greenhouse gas emissions. *Our World in Data.* https://ourworldindata.org/greenhouse-gas-emissions#annual-greenhousegas-emissions-how-much-do-we-emit-each-year; Climate Watch, www.climatewatchdata.org/ghg-emissions?breakBy=gas&chartType=line&end_year=2019&gases=all-ghgCco2&sectors=total-including-lucf&start_ye

Scheffen M, Marchal DG, Beneyton T, Schuller SK, Klose M, Diehl C, Lehmann J, Pfister P, Carrillo M, He H, et al. 2021. A new-to-nature carboxylation module to improve natural and synthetic CO_2 fixation. *Nat Catal* **4:** 105–115. doi:10.1038/s41929-020-00557-y

Sears JE, Boger DL. 2015. Total synthesis of vinblastine, related natural products, and key analogues and development of inspired methodology suitable for the systematic study of their structure–function properties. *Acc Chem Res* **48:** 653–662. doi:10.1021/ar500400w

Sen T, Samanta SK. 2015. Medicinal plants, human health and biodiversity: a broad review. *Adv Biochem Eng Biotechnol* **147:** 59–110. doi:10.1007/10_2014_273

Shine KPF, Fuglestvedt JS, Hailemariam K, Stuber N. 2005. Alternatives to the global warming potential for comparing climate impacts of emissions of greenhouse gases. *Clim Change* **68:** 281–302. doi:10.1007/s10584-005-1146-9

Veeresham C. 2012. Natural products derived from plants as a source of drugs. *J Adv Pharm Technol Res* **3:** 200–201. doi:10.4103/2231-4040.104709

Optimizing Crop Plant Stomatal Density to Mitigate and Adapt to Climate Change

Julie Gray[1,2] and Jessica Dunn[1,2]

[1]Plants, Photosynthesis and Soil, School of Biosciences; [2]Institute for Sustainable Food, University of Sheffield, Sheffield S10 2TN, United Kingdom

Correspondence: j.e.gray@sheffield.ac.uk

Plants take up carbon dioxide, and lose water, through pores on their leaf surfaces called stomata. We have a good understanding of the biochemical signals that control the production of stomata, and over the past decade, these have been manipulated to produce crops with fewer stomata. Crops with abnormally low stomatal densities require less water to produce the same yield and have enhanced drought tolerance. These "water-saver" crops also have improved salinity tolerance and are expected to have increased resistance to some diseases. We calculate that the widespread adoption of water-saver crops could lead to reductions in greenhouse gas emissions equivalent to a maximum of 0.5 $GtCO_2$/yr and thus could help to mitigate the impacts of climate change on agriculture and food security through protecting yields in stressful environments and requiring fewer inputs.

GREENHOUSE GAS EMISSIONS AND CLIMATE CHANGE

Climate change models suggest that we face a warmer future with insufficient water for agriculture (World Bank Group, documents1.world bank.org/curated/en/875921614166983369/pdf/Water-in-Agriculture-Towards-Sustainable-Agriculture.pdf). Increases in greenhouse gas (GHG) emissions have already been linked to a rise in average global temperatures of ~1.1°C, and it appears unlikely that humanity will be able to meet the target of limiting warming to 1.5°C above preindustrial levels set in the COP21 Paris Agreement of 2015 (World Meteorological Organization, public.wmo.int/en/our-mandate/climate/wmo-statement-state-of-global-climate). It is likely that we shall exceed the 1.5°C target, and this could cause an amplification of global warming and make a return to preindustrialized temperatures far more challenging (Intergovernmental Panel on Climate Change [IPCC] 2022). Because agriculture represents one of the major sources of anthropogenic GHG emissions, humanity urgently needs to develop agricultural systems that produce lower levels of GHG emissions by using our precious water, nutrient, and land resources more efficiently. We must also protect against the wasteful GHG emissions associated with yield losses, by adapting crops to be more resilient to the additional stresses imposed by the changing climate.

CLIMATE CHANGE DIMINISHES WATER AVAILABILITY FOR AGRICULTURE

The warmer climate holds more water vapor in the atmosphere, and this is causing the drying of our lakes and rivers, reducing moisture in our soils, and making rainfall less predictable. In combination with altered water flows, this lack of water availability is already affecting productivity in many agricultural areas (FAO 2021a). Indeed, drought causes the single greatest loss of agricultural production accounting for >34% of the losses of crop and livestock production in low- and middle-income countries (FAO 2021b). This problem is expected to get worse in many crop-growing areas of the world, with more frequent and severe drought and heat periods being predicted (IPCC 2022). Although global heating and increasing climate instability are predicted to cause some areas to become wetter (because of severe rainfall events and rising sea levels), most experts predict that the global climate will continue to become hotter and drier. This is expected to intensify land desertification and precipitate more widespread famines and fires (FAO 2021b). When we also consider the requirement to increase agricultural outputs to feed our growing population, the need to create crops that are more resilient to environmental stresses becomes ever more critical (Beddington 2010).

We will not be able to rely on increased levels of irrigation to maintain crop productivity as water availability for crops becomes more limited. Already perhaps 70% of freshwater is used for agriculture, and irrigation pumping uses 6% of global electricity (World Bank Group, documents1 .worldbank.org/curated/en/875921614166983369/ pdf/Water-in-Agriculture-Towards-Sustainable-Agriculture.pdf). As water resources diminish under climate change, we will not be able to irrigate crops to the same extent, and we will need to grow crops that require less water. The use of crops that require less water could provide additional benefits, as irrigation can create further problems that impact negatively on long-term productivity, such as increased soil salinity and toxicity.

The current challenge is to produce crops that are both resilient to periods of water scarcity and able to use the available water more efficiently. These are known as drought resilience and water use efficiency, respectively.

STOMATA CONTROL PLANT WATER LOSS

Several biotechnological routes have been proposed to create crops that are more resilient to drought, including up-regulating stress-related gene transcription or improving root architecture. Such manipulations could help to mitigate yield reductions under climate change (Martignago et al. 2020). One potential route to both increasing drought tolerance and reducing crop water requirements that we have recently proposed lies in reducing the number of water-permeable pores known as stomata (Bertolino et al. 2019; Buckley et al. 2020). Stomata are microscopic adjustable pores on the surface of leaves that allow plants to take up carbon dioxide (Fig. 1). The primary role of stomata is to facilitate photosynthesis, which underpins all crop yields. However, for each CO_2 molecule fixed, hundreds of water molecules are lost through the open pores, and below we discuss how this ratio of CO_2 uptake to water loss could be better optimized for current and future climate conditions. The water that is taken up by roots moves through the plant in the transpiration stream and is rapidly, and almost completely, lost through the stomata on the leaves. This water is not entirely wasted, though, as the transpiration stream pulls vital nutrients with it toward the growing tips and seeds, as well as cooling the plant tissues. However, any toxins or salts in the soil may also move with the transpiration stream and accumulate in the resulting crops, causing a drop in yields or compromising food safety. In addition, stomata can be exploited by many fungal and bacterial crop pathogens that use these pores as convenient entry points into the interior of the plant. Thus, although stomata are vital for crop photosynthesis, there are costs to the plant of having excessive stomatal capacity including a reduction in water use efficiency (WUE) (i.e., the ratio of carbon gained to water lost) and potentially an increased susceptibility to drought, salinity, and disease. To ensure that these negative attributes of stomata do not outweigh the benefits, plants

Cite this article as *Cold Spring Harb Perspect Biol* doi: 10.1101/cshperspect.a041672

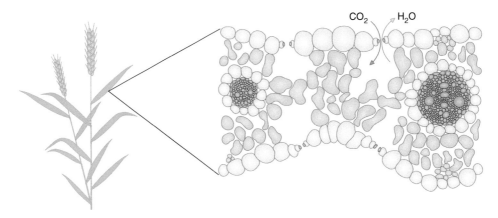

Figure 1. Diagram of a wheat plant showing a cross section through a leaf to reveal the cellular patterning. Stomata are present in the epidermal layers and consist of pairs of guard cells and subsidiary cells (purple and blue). These minute adjustable pores regulate the flow of gases between the atmosphere and the interior of the leaf. When the pores are open in the light, carbon dioxide enters the leaf and is fixed by photosynthesis. At the same time, water vapor is lost to the atmosphere through the transpiration stream.

have evolved the ability to close their stomata at night when there is no light for photosynthesis and also to shut in response to drought or pathogens. Efficient stomatal closure mechanisms allow plants to conserve water and improve their WUE (Lawson and Vialet-Chabrand 2019). However, stomata cannot completely close, and water loss continues at a lower level even in the dark or under drought conditions. Thus, the number and size of stomatal pores is a major determinant of the achievable minimum (and maximum) level of crop WUE.

PLANTS WITH FEWER STOMATA COULD BE BETTER SUITED TO FUTURE CLIMATES AND CO$_2$ LEVELS

Recent improvements in our understanding of how stomata are produced have enabled novel ways to generate crops that both use water more efficiently and are drought-tolerant. In brief, studies using the model plant *Arabidopsis* show that early in leaf development, biochemical signals are secreted from stomatal precursor cells. These signals activate receptors on the surface of surrounding epidermal cells, preventing them from entering the stomatal development program (by promoting degradation of the transcriptional regulator that is responsible for initiating stomatal development). At the same

time, antagonizing signals are secreted from the mesophyll cells of the leaf interior, and these promote stomatal development in the epidermis by competing for and blocking the same receptors (Zoulias et al. 2018; Lee and Bergmann 2019; Torii 2021). Thus, an intricate balance of activating and inhibiting developmental signals determines the density of stomata that result on the mature leaf. These secreted peptide signals are encoded by members of the *epidermal patterning factor (EPF)* gene family, which includes *EPF* and *EPF-like (EPFL)* members. This family includes several members that have directly opposing effects, acting to either inhibit or promote the number of stomata that eventually form on the leaves (e.g., Hara et al. 2007; Hunt and Gray 2009; Sugano et al. 2010).

Plants lacking a particular EPF (called EPF2 in *Arabidopsis)* produce more stomata, whereas those genetically engineered to overexpress this signal produce very low densities of stomata (Hunt and Gray 2009). Remarkably, EPF-overexpressing *Arabidopsis* plants with only 20% of the normal number of stomata appear to grow normally and have enhanced soil water conservation and WUE (i.e., an increased proportion of CO$_2$ molecules fixed per water molecules lost) and increased levels of drought tolerance. We have called these plants with reduced stomatal density "water-saver" plants because of their ability to conserve soil water and avoid drought stress, without showing any significant reduction in photosynthesis or nutrient uptake (Doheny-Adams et al. 2012; Hepworth et al. 2015; Harrison et al. 2020). As described below, the water-saver plants also show a degree of pathogen tolerance because of the reduced number of stomatal entry points (Dutton et al. 2019). We have also discovered that similar EPFs and EPFLs exist across remarkably diverse groups of land plants, including important cereal crop species (Hepworth et al. 2015). In an effort to create crops that require less water, water-saver crops have also recently been created through reducing stomatal density by overexpressing similar crop EPFs.

The discovery of these biochemical signals has allowed us to test whether crop plants really need so many stomata, and what would happen if we reduced the density of stomata on crop leaves (Harrison et al. 2020). It is timely to ask these questions because carbon dioxide levels are currently rising so rapidly that plants are living at much higher levels of CO_2 than they evolved to grow at. About 100 years ago, CO_2 levels were ~280 ppm. They have now passed 400 ppm and are still rising (National Oceanic and Atmospheric Administration, www.noaa .gov/news-release/carbon-dioxide-now-more-than-50-higher-than-pre-industrial-levels). Plants should need a much lower capacity for CO_2 uptake when CO_2 levels are so high. However, the increases in CO_2 emissions have happened too quickly for evolutionary change to occur, and it is calculated that most plants are currently operating at about only 10%–20% of their stomatal capacity (Franks et al. 2015). Many plant species can make some developmental adaptions to high CO_2 levels, and these often include making small reductions in stomatal development. Unfortunately, many domesticated crop species, including rice, are unable to reduce the number of stomata that they produce at high CO_2 (Rowland-Bamford et al. 1990). This suggests that at current CO_2 levels, if crops could be produced that have fewer stomata, these should still produce the same yield but require much less water to do so. They should also accumulate lower levels of toxins such as salt and arsenic in their tissues when grown in contaminated soils and be less susceptible to diseases caused by pathogens that infect through stomata. Crop species that can reduce their numbers of stomata at high CO_2 can only make small adjustments. We, therefore, require a biotechnological approach that can substantially reduce stomatal numbers to realize the improvements in drought resilience, WUE, and salinity and pathogen tolerances discussed below.

CROPS WITH FEWER STOMATA REQUIRE LESS WATER AND ARE MORE DROUGHT-TOLERANT

Some of our most important crops are cereals, and their cultivation is particularly water-intensive, with some rice and wheat agricultural systems requiring thousands of liters of water to produce a single kilogram of grain. Homologs of *EPFs* have now been identified in barley, wheat, and rice genomes, and have been overexpressed in these crops (Hughes et al. 2017; Caine et al. 2019; Dunn et al. 2019; Lu et al. 2019; Mohammed et al. 2019). In all three crops, as in *Arabidopsis*, EPF overexpression resulted in substantial reductions in stomatal density (Fig. 2). It has now been demonstrated for each of these three cereal species, that the EPF-engineered crops with approximately half of the normal number of stomata on their leaves have enhanced WUE with little or no detriment to photosynthesis and, most importantly, no significant yield penalty. For example, when rice plants with low stomatal density were cultivated under well-watered conditions, they conserved more water in the soil and required 42% less water, yet they still produced the same amount of grain as the control plants (Fig. 3). When the water-saver rice plants were subjected to a drought treatment they were more able to maintain their photosynthesis and growth and went on to produce ~27% more grain yield than their controls in the same conditions (Fig. 3; Caine et al. 2019). Their conservative use of water when it was plentiful allowed the water-saver rice to preserve moisture levels in the soil, enabling drought stress avoidance when water became scarce.

Rice cultivation probably produces the largest human-caused emissions of methane, which

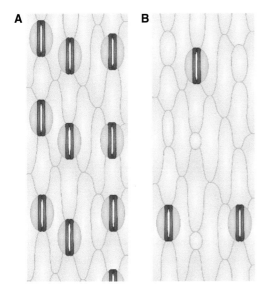

Figure 2. The overexpression of an epidermal patterning factor (EPF) in genetically modified wheat, barley, or rice plants leads to a reduction in the density of stomata that form in the leaf epidermis. These EPF-overexpressing plants with fewer stomata lose less water and they have, therefore, been named water-saver. Low stomatal density plants have enhanced water use efficiency, are drought-tolerant, and are better able to survive in saline conditions. The figure shows cell patterning of the leaf epidermis of (A) control plants with the normal complement of stomata, and (B) EPF-overexpressing plants with fewer stomata.

is a much more potent but shorter-lived GHG than carbon dioxide. Anoxic soil conditions of paddy fields in Asia encourage methanogenesis (Kumar and Ladha 2011). Altering rice agricultural practices to move toward aerobic direct or dry seeded rice or alternate wetting and drying would be preferable, and, where appropriate, this could reduce methane emissions by up to 75% and facilitate more intensive rice production (Pathak 2023). However, flooding helps to manage weed problems and yields of direct-seeded rice have been variable. Growers are, therefore, reluctant to change their agricultural practices. The availability of drought-tolerant water-saver cultivars could help to encourage the much-needed move away from flooded paddy fields as direct-seeded rice is so much more vulnerable to drought stress.

PLANTS WITH FEWER STOMATA HAVE ENHANCED PATHOGEN AND SALINITY TOLERANCE

Experiments with *Arabidopsis* mutants and *Pseudomonas syringae*, a bacterium that gains access to the plant through stomata and causes bacterial speck disease (Fig. 4), have shown that plants with reduced stomatal density have significantly reduced pathogen loads following infection. Disease-related symptoms were less severe for several days after inoculation and photosynthetic efficiency was maintained at higher levels in the *Arabidopsis* plants with fewer stomata (Dutton et al. 2019). This suggests that a decrease in stomatal pathogen entry points could give

Figure 3. Water-saver rice uses less water and seed yields are more resilient to drought conditions. (A) When plants are well supplied with water, the volume of water taken up by an epidermal patterning factor (EPF)-overexpressing rice plant (*right*) with 49% of the normal stomatal density on the first true leaf is 42% less than the volume of water taken up by a control plant (*left*) with the normal complement of stomata over the same 1-wk period (28–35 d postgermination). (B) The grain yield from an EPF-overexpressing rice plant is 27% higher than in a control plant following the imposition of a 3-d severe drought period during the flowering period. (Figure created from data in Caine et al. 2019.)

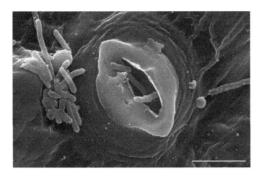

Figure 4. Disease-causing pathogens can enter through stomata. A scanning electron microscope image of *Pseudomonas syringae* cells on the epidermal surface of *Arabidopsis thaliana*. Scale bar, 5 μm. (Figure provided by L. Hunt and C. Dutton.)

plants additional time to mount an appropriate defense response; this could slow the spread of an infection and help to protect yields. Thus, low stomatal water-saver barley and wheat plants could have significant tolerance to fungal pathogens that enter through stomata and cause diseases such as brown and yellow rust. Rice does not have a rust-causing pathogen but is susceptible to several bacterial pathogens that gain entry via stomata. The most important of these are *Xanthomonas oryzae* pv. *oryzae* and *Xanthomonas oryzae* pv. *oryzicola*, which cause leaf blight and leaf streak diseases, which both constrain rice productivity (Niño-Liu et al. 2006). Because many important fungal and bacterial crop pathogens cause substantial losses by infection through stomatal pores, it might be expected that water-saver crops manipulated to have low stomatal density would show a degree of disease resistance that could help to protect against yield losses and reduce the levels of pesticide application required. Again, this research area has exciting potential but requires more attention.

The growth and survival of most plants, and particularly crop species, are severely affected by high levels of salts in soils. The area of agricultural land affected by salinity, and therefore lost to food production, is continually expanding mostly because of rising sea levels (which accompany global warming) and water extraction and unsustainable irrigation practices. A World Bank working group has calculated that the agricultural losses to salinity each year are equivalent to the amount of food required to feed 170 million people (Russ et al. 2020). This means that year on year salinity causes the loss of enough farming land to feed half of the population of the United States. We urgently need to slow this loss in agricultural capacity by developing crops that require less irrigation and are better able to tolerate saline conditions. Recently, in addition to having increased drought and pathogen tolerances as described above, water-saver rice has been shown to be able to survive periods of salinity (Caine et al. 2023). Perhaps because of its reduced levels of water loss, this rice takes up and accumulates less salt and has an enhanced survival rate when exposed to salinity treatments that are normally lethal to rice plants (20 mM or 50 mM NaCl applied over several weeks). The water-saver rice accumulates less than half the amount of sodium in its leaves and outperforms traditionally bred stress-tolerant cultivars when grown in saline conditions (Caine et al. 2023).

WATER-SAVING RICE DOES NOT SHOW ENHANCED SUSCEPTIBILITY TO HEAT STRESS

As our climate warms, it is predicted that in the near future heat stress will cause a more severe challenge to agricultural production than drought stress (Semenov and Shewry 2011). Cereal yields are susceptible to heat stress, particularly during reproductive and grain-filling stages. Over the next 25 years, it is expected that one-quarter of the rice-growing areas will experience yield losses from high temperatures (Jagadish et al. 2015). Perhaps worryingly, the water-saver rice with reduced stomatal density showed slightly increased leaf surface temperatures when well supplied with water (Caine et al. 2019). This is presumed to be due to a reduction in their transpirational water loss and an associated lower level of evaporative cooling. As rising temperatures are expected to be detrimental to rice productivity, it is important to ascertain whether the water-saver plants have increased susceptibility to heat stress.

To investigate this, the temperature of water-saver rice leaves was monitored while they were

Cite this article as *Cold Spring Harb Perspect Biol* doi: 10.1101/cshperspect.a041672

exposed to heat conditions known to be damaging to rice (35°C or 40°C). Surprisingly, the low stomatal density leaves were able to maintain the same temperature as the control plants (which had the normal complement of stomata). This indicates that water-saver rice has sufficient evaporative cooling capacity to cope with air temperatures up to at least 40°C. Although the plants have fewer stomata, the stomata that they have are functional and they are able to open and close normally. Rather than succumb to overheating, the low-stomatal-density plants stopped saving water. At high temperatures, they opened their stomata wider and were able to cool themselves as efficiently as the control plants (Caine et al. 2019). It, therefore, appears that water-saver crops are not particularly susceptible to rising temperatures, possibly because, as mentioned above, most plants are operating at only a small proportion of their stomatal capacity. Although promising, yields under heat stress are yet to be reported.

MANIPULATING STOMATAL DENSITY COULD REDUCE GHG EMISSIONS

Together, the above results demonstrate that reducing the stomatal density of crops could give several advantages. These crops should be less susceptible to yield loss when exposed to drought, salinity, disease-causing microbes that enter through stomata, or several of these in combination. Furthermore, they should not show a yield deficit when well supplied with water nor an enhanced susceptibility to heat stress. Thus, water-saver crops should be particularly well-suited to the multistress environments that are occurring more frequently in agricultural areas under climate warming. Hence, it is possible that their adoption could help us to adapt to, and reduce, the damaging effects of climate change on agriculture (Bertolino et al. 2019; Buckley et al. 2020). Below we outline calculations to quantify the maximum level of GHG reductions that could be achieved by the adoption of water-saver crops.

Global agriculture and related land use were estimated to cause 9.8 billion tonnes of carbon dioxide equivalent ($GtCO_2e$) emissions in 2018, of which ~10% came from rice cultivation alone (FAO 2020). Above, we outlined how producing crops with reduced stomatal densities could enhance stress tolerances and protect yields and these, combined with improved agricultural practices, could lead to a drop in GHG emissions. Our calculations shown in Table 1 suggest that if (in an optimistic scenario) low-stomatal-density traits were to be fully adopted by growers, the drought-resilient and WUE properties of water-saver rice have the potential to make substantial reductions in agricultural GHG emissions, up to 417.7 million tonnes of carbon dioxide equivalent per year ($MtCO_2e/yr$). Additional reductions of 67.5 $MtCO_2e/yr$ could be achieved from moderate reductions in wider agricultural losses to salinity and to pathogens that enter through stomata (Table 2). Thus, the combined potential reductions from this technology could reach 485.2 $MtCO_2e/yr$, equating to 1.3% of the 36.2 $GtCO_2e/yr$ of total anthropogenic emissions or 4.4% of the 11 $GtCO_2e/yr$ emission re-

Table 1. Maximum potential greenhouse gas emission savings from the adoption of low-stomatal-density rice

Yield protected from drought loss	147.7[a]
Energy use reduction from decreased irrigation	12.5[b]
Shift from paddy to dry/direct-seeded rice	257.5[c]
Total potential reductions	417.7 $MtCO_2e/yr$

Data sources: [a]Rice cultivation accounts for 1.5%–2.5% of the estimated 36.2 $GtCO_2e/yr$ human-induced GHG emissions (LeQuéré et al. 2018). Drought reduces agricultural yields by 34% (FAO 2021b) and low stomatal density rice reduces the impact of drought on yield by 60% (Caine et al. 2019).

[b]Difference between energy inputs for irrigated and rainfed rice in Myanmar is 8139.4 MJ/ha (Soni and Soe 2016). On average, 818.3 lb CO_2 emitted per MWh energy generated or 0.13 kg/MJ (U.S. Environmental Protection Agency, eGRID Power Profiler). Global area of irrigated rice estimated at 118 Mha (75% of total rice area of 158 Mha).

[c]Paddy rice produces an estimated 20–100 Tg CH_4/yr. Better management could reduce methane emissions by 270 $MtCO_2e/yr$ (Working Group III IPCC, Smith et al. 2007). To avoid double accounting, 12.5 Mt savings from reduced irrigation have been deducted.

Table 2. Potential greenhouse gas reductions from improved tolerance to pathogens and salinity through adoption of low-stomatal-density crops

10% Reduction in loss to fungal diseases	56[a]
10% Reduction in fungicide production	0.5[b]
10% Reduction in loss of land to salinity	11[c]
Potential reductions	67.5 MtCO$_2$e/yr

Data sources: [a]Fungal diseases cause an estimated 11% loss in crop yields and agriculture accounts for estimated emissions of 5.1–6.1 GtCO$_2$e/yr (Working Group III of IPCC, Smith et al. 2007).
[b]Pesticide energy input of 1364 MJ/ha corresponds to weighted average GHG emission of 94 kg CO$_2$e/ha arable crop, 27 kg CO$_2$e/ha of which is fungicide (Audsley et al. 2009). Global area of cropland is ~166 Mha (FAO 2021a).
[c]Salinity causes the loss of ~0.02% of global agricultural yield each year (World Bank Policy research working paper, Russ et al. 2020).

ductions that are predicted to be required to stay within the 1.5°C global warming target. Wholesale adoption of new agricultural technology is, of course, an overoptimistic scenario, and the potential benefits derived from the application of this technology would vary between agricultural systems, climatic regions, and crop species. Nonetheless, field trials combined with further assessment of these calculations could enable future targeting of research and development in which the maximum GHG reductions could be achieved. Seed-based technologies are relatively low-cost to implement and the impact on reduced GHG emissions, which has been estimated here based on a range of existing evidence, is potentially substantial.

ROUTES TO THE FIELD

Genetic tools are currently available that could rapidly produce crop varieties that are better able to cope with abiotic and biotic stresses and that we calculate should have reduced GHG emissions associated with their production. Through simply overexpressing an EPF, we can reduce stomatal density, decrease crop water use, conserve soil water, and improve resilience to drought and salinity and perhaps some pathogens. The experiments to date have focused on making substantial reductions in crop stomatal density (i.e., 50% or greater) to provide "proof of principle" that a reduction in stomatal density can improve stress tolerances (Hughes et al. 2017; Caine et al. 2019; Dunn et al. 2019). Further work now needs to be carried out to determine the particular stomatal densities required to protect yields and to optimize the reductions in GHG emissions that

might be achievable for different crop species in differing agricultural systems and environments —in particular, for direct-seeded rice systems.

Existing legislation and public opposition currently ensure that crops overexpressing an EPF would be expensive and slow to implement. There is still substantial opposition to genetically engineered crops and, because of the urgency of the climate crisis, it is critical that other potential routes are explored so that we can create similar water-saver varieties in perhaps more publicly acceptable ways, as soon as possible. Considerable efforts have already begun to achieve this via different methodologies. For example, rice with substantially reduced stomatal density has been produced using gene editing to target and knock out an antagonistic EPFL biochemical signal that normally promotes stomatal development (Yin et al. 2017). Furthermore, through the screening of a fast-neutron mutagenized rice population, several stomatal density and size traits have recently been identified that are readily accessible for the conventional breeding of improved water-saver crops (Pitaloka et al. 2022).

Most of the results discussed in the above sections have been taken from trials carried out in greenhouses or controlled environment chambers (because of the complications and costs arising from current restrictions on the release of genetically modified [GM] crops). Although these results are encouraging in terms of stress tolerances and yields, it is now important that low-stomatal-density water-saver crops are properly assessed in the field under different climatic conditions. We also need to investigate how this trait will perform in combination with other known genetic traits associated with drought and disease resistance and whether stacking of multiple crop improvement traits could further protect crop yields and enhance GHG emission savings. The recent availability of non-GM varieties with low stomatal density should allow field trials in the near future.

In conclusion, research across several major cereal species has demonstrated that stomatal development can be substantially reduced, and this could bring about major savings in crop water use without impacting on seed yield. Indeed, water-saver crops have the potential to yield better than conventional varieties under challenging environments. They could also bring about a reduction in water pumping for irrigation and may help to facilitate a move away from highly methanogenic paddy field agriculture. Thus, crops with low stomatal densities have the potential to improve the resilience of crops to the ongoing effects of climate change while perhaps reducing the level of GHG emissions produced by agriculture, thereby helping humanity to both adapt to and mitigate future climate change.

ACKNOWLEDGMENTS

J.G. is funded by a Royal Society Leverhulme Trust Senior Research Fellowship and J.D. by a Biotechnology and Biological Sciences Research Council (BBSRC) White Rose studentship and a Postdoctoral Fellowship from the Institute for Sustainable Food. This article has been made freely available online by generous financial support from the Bill and Melinda Gates Foundation. We particularly want to acknowledge the support and encouragement of Rodger Voorhies, president, Global Growth and Opportunity Fund.

REFERENCES

Audsley E, Stacey K, Parsons DJ, Williams A. 2009. *Estimation of the greenhouse gas emissions from agricultural pesticide manufacture and use.* Cranfield University, Bedford, UK. doi:10.13140/RG.2.1.5095.3122

Beddington J. 2010. Global food and farming futures. *Philos Trans R Soc Lond B Biol Sci* **365**: 2767. doi:10.1098/rstb.2010.0181

Bertolino LT, Caine RS, Gray JE. 2019. Impact of stomatal density and morphology on water-use efficiency in a changing world. *Front Plant Sci* **10**: 225. doi:10.3389/fpls.2019.00225

Buckley CR, Caine RS, Gray JE. 2020. Pores for thought: can genetic manipulation of stomatal density protect future rice yields? *Front Plant Sci* **10**: 1783. doi:10.3389/fpls.2019.01783

Caine RS, Yin X, Sloan J, Harrison EL, Mohammed U, Fulton T, Biswal AK, Dionora J, Chater CC, Coe RA, et al. 2019. Rice with reduced stomatal density conserves water and has improved drought tolerance under future climate conditions. *New Phytol* **221**: 371–384. doi:10.1111/nph.15344

Caine RS, Harrison EL, Sloan J, Flis P, Fischer S, Khan M, Nguyen P, Nguyen TL, Gray JE, Croft H. 2023. The influences of stomatal size and density on rice abiotic stress resilience. *New Phytol* **237**: 2180–2195. doi:10.1111/nph.18704

Doheny-Adams T, Hunt L, Franks PJ, Beerling DJ, Gray JE. 2012. Genetic manipulation of stomatal density influences stomatal size, plant growth and tolerance to restricted water supply across a growth carbon dioxide gradient. *Philos Trans R Soc Lond B Biol Sci* **367**: 547–555. doi:10.1098/rstb.2011.0272

Dunn J, Hunt L, Afsharinafar M, Meselmani MA, Mitchell A, Howells R, Wallington E, Fleming AJ, Gray JE. 2019. Reduced stomatal density in bread wheat leads to increased water-use efficiency. *J Exp Bot* **70**: 4737–4748. doi:10.1093/jxb/erz248

Dutton C, Hõrak H, Hepworth C, Mitchell A, Ton J, Hunt L, Gray JE. 2019. Bacterial infection systemically suppresses stomatal density. *Plant Cell Environ* **42**: 2411–2421. doi:10.1111/pce.13570

FAO. 2020. Emissions due to agriculture. Global, regional and country trends 2000–2018. FAOSTAT Analytical Brief Series No. 18. Rome, Italy.

FAO. 2021a. World food and agriculture—statistical yearbook 2021. Rome, Italy. doi:10.4060/cb4477en

FAO. 2021b. The impact of disasters and crises on agriculture and food security: 2021. Rome, Italy. doi:10.4060/cb3673en

Franks PJ, W Doheny-Adams T, Britton-Harper ZJ, Gray JE. 2015. Increasing water-use efficiency directly through genetic manipulation of stomatal density. *New Phytol* **207**: 188–195. doi:10.1111/nph.13347

Hara K, Kajita R, Torii KU, Bergmann DC, Kakimoto T. 2007. The secretory peptide gene EPF1 enforces the stomatal one-cell-spacing rule. *Genes Dev* **21**: 1720–1725. doi:10.1101/gad.1550707

Harrison EL, Arce Cubas L, Gray JE, Hepworth C. 2020. The influence of stomatal morphology and distribution on photosynthetic gas exchange. *Plant J* **101**: 768–779. doi:10.1111/tpj.14560

Hepworth C, Doheny-Adams T, Hunt L, Cameron DD, Gray JE. 2015. Manipulating stomatal density enhances drought tolerance without deleterious effect on nutrient uptake. *New Phytol* **208**: 336–341. doi:10.1111/nph.13598

Hughes J, Hepworth C, Dutton C, Dunn JA, Hunt L, Stephens J, Waugh R, Cameron DD, Gray JE. 2017. Reducing stomatal density in barley improves drought tolerance without impacting on yield. *Plant Physiol* **174**: 776–787. doi:10.1104/pp.16.01844

Hunt L, Gray JE. 2009. The signaling peptide EPF2 controls asymmetric cell divisions during stomatal development. *Curr Biol* **19**: 864–869. doi:10.1016/j.cub.2009.03.069

Intergovernmental Panel on Climate Change (IPCC). 2022. Summary for policymakers. Climate change 2022: impacts, adaptation and vulnerability. Contribution of Working Group II to the Sixth Assessment Report of the Intergovernmental Panel on Climate Change. Cambridge University Press, Cambridge, UK. doi:10.1017/9781009325844.001

Jagadish SV, Murty MV, Quick WP. 2015. Rice responses to rising temperatures—challenges, perspectives and future directions. *Plant Cell Environ* **38**: 1686–1698. doi:10.1111/pce.12430

Kumar V, Ladha JK. 2011. Direct seeding of rice: recent developments and future research needs. *Adv Agron* **111**: 297–413. doi:10.1016/B978-0-12-387689-8.00001-1

Lawson T, Vialet-Chabrand S. 2019. Speedy stomata, photosynthesis and plant water use efficiency. *New Phytol* **221**: 93–98. doi:10.1111/nph.15330

Lee LR, Bergmann DC. 2019. The plant stomatal lineage at a glance. *J Cell Sci* **132**: jcs228551. doi:10.1242/jcs.228551

LeQuéré C, Andrew RM, Friedlingstein P, Sitch S, Pongratz J, Manning AC, Korsbakken J, Peters GP, Canadell JG, Jackson RB, et al. 2018. Global carbon budget 2017. *Earth Syst Sci Data* **10**: 405–448. doi:10.5194/essd-10-405-2018

Lu J, He J, Zhou X, Zhong J, Li J, Liang YK. 2019. Homologous genes of epidermal patterning factor regulate stomatal development in rice. *J Plant Physiol* **234–235**: 18–27. doi:10.1016/j.jplph.2019.01.010

Martignago D, Rico-Medina A, Blasco-Escámez D, Fontanet-Manzaneque JB, Caño-Delgado AI. 2020. Drought resistance by engineering plant tissue-specific responses. *Front Plant Sci* **10**: 1676. doi:10.3389/fpls.2019.01676

Mohammed U, Caine RS, Atkinson JA, Harrison EL, Wells D, Chater CC, Gray JE, Swarup R, Murchie EH. 2019. Rice plants overexpressing OsEPF1 show reduced stomatal density and increased root cortical aerenchyma formation. *Sci Rep* **91**: 1–13. doi:10.1038/s41598-019-41922-7. doi:10.1038/s41598-019-51402-7

Niño-Liu DO, Ronald PC, Bogdanove AJ. 2006. *Xanthomonas oryzae* pathovars: model pathogens of a model crop. *Mol Plant Pathol* **7**: 303–324. doi:10.1111/j.1364-3703.2006.00344.x

Pathak H. 2023. Impact, adaptation, and mitigation of climate change in Indian agriculture. *Environ Monit Assess* **195**: 52. doi:10.1007/s10661-022-10537-3

Pitaloka MK, Caine RS, Hepworth C, Harrison EL, Sloan J, Chutteang C, Phunthong C, Nongngok R, Toojinda T, Ruengphayak S, et al. 2022. Induced genetic variations in stomatal density and size of rice strongly affects water use efficiency and responses to drought stresses. *Front Plant Sci* **13**: 801706. doi:10.3389/fpls.2022.801706

Roser M, Ritchie H. 2013. Food supply. Our world in data. https://ourworldindata.org/food-supply

Rowland-Bamford AJ, Nordenbrock C, Baker JT, Bowes G, Hartwell A Jr. 1990. Changes in stomatal density in rice grown under various CO_2 regimes with natural solar irradiance. *Environ Exp Bot* **30**: 175–180. doi:10.1016/0098-8472(90)90062-9

Russ J, Zaveri E, Damania R, Desbureaux S, Escurra J, Rodella A. 2020. Salt of the earth quantifying the impact of water salinity on global agricultural productivity. Policy Research Working Paper 9144. World Bank Group. doi:10.1596/1813-9450-9144

Semenov MA, Shewry PR. 2011. Modelling predicts that heat stress, not drought, will increase vulnerability of wheat in Europe. *Sci Rep* **1**: 66. doi:10.1038/srep00066

Smith PD, Martino Z, Cai D, Gwary H, Janzen P, Kumar B, McCarl S, Ogle F, O'Mara C, Rice B, et al. 2007. Agriculture. Climate change 2007: Mitigation. Contribution of Working Group III to the Fourth Assessment Report of the Intergovernmental Panel on Climate Change. Cambridge University Press, Cambridge.

Soni P, Soe MN. 2016. Energy balance and energy economic analyses of rice production systems in Ayeyarwaddy Region of Myanmar. *Energy Efficiency* **9**: 223–237. doi:10.1007/s12053-015-9359-x

Sugano SS, Shimada T, Imai Y, Okawa K, Tamai A, Mori M, Hara-Nishimura I. 2010. Stomagen positively regulates stomatal density in *Arabidopsis*. *Nature* **463**: 241–244. doi:10.1038/nature08682

Torii KU. 2021. Stomatal development in the context of epidermal tissues. *Ann Bot* **128**: 137–148. doi:10.1093/aob/mcab052

Yin X, Biswal AK, Dionora J, Perdigon KM, Balahadia CP, Mazumdar S, Chater C, Lin HC, Coe RA, Kretzschmar T, et al. 2017. CRISPR-Cas9 and CRISPR-Cpf1 mediated targeting of a stomatal developmental gene EPFL9 in rice. *Plant Cell Rep* **36**: 745–757. doi:10.1007/s00299-017-2118-z

Zoulias N, Harrison EL, Casson SA, Gray JE. 2018. Molecular control of stomatal development. *Biochem J* **475**: 441–454. doi:10.1042/BCJ20170413

Direct Methane Removal from Air by Aerobic Methanotrophs

Mary E. Lidstrom

Department of Chemical Engineering, Department of Microbiology, University of Washington, Seattle, Washington 98195, USA

Correspondence: lidstrom@uw.edu

The rapid pace of climate change has created great urgency for short term mitigation strategies. Appropriately, the long-term target for intervening in global warming is CO_2, but experts suggest that methane should be a key short-term target. Methane has a warming impact 34 times greater than CO_2 on a 100-year timescale, and 86 times greater on a 20-year timescale, and its short half-life in the atmosphere provides the opportunity for near-term positive climate impacts. One approach to removing methane is the use of bacteria for which methane is their sole carbon and energy source (methanotrophs). Such bacteria convert methane to CO_2 and biomass, a potentially value-added product and co-benefit. If air above emissions sites with elevated methane is targeted, technology harnessing the aerobic methanotrophs has the potential to become economically viable and environmentally sound. This article discusses challenges and opportunities for using aerobic methanotrophs for methane removal from air, including the avoidance of increased N_2O emissions.

METHANE AND CLIMATE CHANGE

The rapid pace of climate change has created great urgency for mitigation measures (Intergovernmental Panel on Climate Change [IPCC] 2021). CO_2 continues to be a major target, but it is now broadly accepted that a second prime target for slowing global warming by 2050 is the greenhouse gas methane (Smith et al. 2020; Intergovernmental Panel on Climate Change [IPCC] 2021; Jackson et al. 2021; Ocko et al. 2021; Sun et al. 2021; Dreyfus et al. 2022). Methane has a warming impact 34 times greater than CO_2 on a 100-yr timescale, and 86 times greater on a 20-yr timescale (Abernethy et al. 2021; Intergovernmental Panel on Climate Change [IPCC] 2021).

In addition, its relatively short half-life in the atmosphere ($\sim$10 yr) provides the opportunity for near-term climate mitigation (Smith et al. 2020; Abernethy et al. 2021; Intergovernmental Panel on Climate Change [IPCC] 2021; Ocko et al. 2021; Sun et al. 2021). The United States has recently joined more than 100 countries in the Global Methane Pledge to reduce methane emissions by 2030 (U.S. Department of State 2021). Recent projections predict that a significant impact can be made by 2050. In one modeling study, 0.3 petagrams (gigatonnes [Gt]) of methane removed by 2050 is predicted to slow global warming by 0.22°C (Ocko et al. 2021), whereas in a second study, 1 petagram (Gt) of methane removal is predicted to slow global warming by 0.21°C (Abernethy et al. 2021). Decreases of this magnitude are predicted to have a major positive impact on our climate future (Abernethy et al. 2021; Ocko et al. 2021), especially when combined with other mitigation strategies (Sun et al. 2021).

EMISSION STRATEGIES FOR DECREASING METHANE: NEED FOR METHANE REMOVAL APPROACHES

Most solutions being proposed for decreasing methane in the atmosphere are focused on decreasing emissions (Ocko et al. 2021), and these are important goals. However, not all methane emissions are amenable to reduction, and it has been argued that emission-reduction strategies must be augmented by methane removal from air to slow global warming by 2050 (Jackson et al. 2021; Warszawski et al. 2021; Nisbet-Jones et al. 2022). In addition, many emission sources will leak significant methane, even after mitigation efforts, because most emission mitigations are only partially effective. For example, Swedish landfills in which methane is harvested still emit more than half of the produced methane to the air (Börjesson et al. 2009). One advantage of methane removal technology is that it can be deployed at any emission site and does not require customization depending on the emission type (e.g., landfills vs. sewage treatment plants vs. rice paddies), which will streamline development and implementation. Finally, it is now becoming clear that many methane emission reduction strategies have the potential to result in increased nitrous oxide (N_2O) emissions because of the complex interplay in microbial communities between methane consumption, denitrification (use of oxidized nitrogen compounds as an electron acceptor), and nitrification (use of ammonia as an energy source) and the generation of N_2O as a by-product of the latter two processes (Stein 2020; Chang et al. 2021; Kang et al. 2021; Nisbet-Jones et al. 2022). Because N_2O is 10 times more potent than methane on a 100-year timescale, if N_2O increases at a rate of only 10% of the methane reduction, they cancel each other out in terms of impact. A recent study with rice paddy soil has shown that N_2O increases in response to increased methane consumption as a result of competition for copper (Chang et al. 2021). Analysis of the results in that study suggests that the N_2O produced is roughly 10% of the methane consumed. In studies of methane and N_2O generation from landfills in which aeration was used to stimulate methane consumption and other degradation rates, N_2O production increased (Nag et al. 2016) and N_2O emissions greater than methane were noted from the working face of a German landfill (Harborth et al. 2013). Because N_2O is rarely measured in methane emission mitigation studies, the extent of N_2O increase is uncertain, but it is possible that many measures taken to reduce methane emissions have the potential to simultaneously increase N_2O emissions. It may be possible to mitigate some of this increase—for instance with nitrification inhibitors (Ocko et al. 2021)—but significant N_2O emissions result from denitrification (Stein 2020; Chang et al. 2021), and denitrification inhibitors have so far shown mixed results (Ye et al. 2022).

METHANE REMOVAL CHALLENGES

Two major challenges for methane removal from air are concentration and scale. Both must be addressed when considering a specific methane mitigation technology to understand potential feasibility and impact.

Concentration

The current concentration of methane in air is very low, at 1.9 ppm, which is ~2 nM dissolved methane in water at 25°C. Certain types of bacteria, called methanotrophs, are known that subsist on methane (see below), and a subset of these strains can consume methane at ambient levels. However, the rates at 1.9 ppm are extremely slow (on the order of $\mu g\ CH_4\ (g\ cells)^{-1}\ h^{-1}$) and not predicted to sustain growth long-term (Kneif and Dunfield 2005; Yoon et al. 2009; Tveit et al. 2019). Alternatively, many methanotrophs consume methane at significant rates at 100–1000 ppm (0.01%–0.1%) (Kneif and Dunfield 2005), and many types of methane emission sites have been shown to have methane in the air directly above or beside the sites at concentrations of 100

Table 1. Examples of documented methane concentrations in air at emission sites

Emission site type	Examples of measured concentration in air (ppm)	References
Landfills active	4000–8000 100–800	Carmen and Vincent 1998; Börjesson et al. 2009
Landfills inactive	1000–3000	Carmen and Vincent 1998
Coal mines active	10,000–50,000	Limbri et al. 2014
Coal mines inactive	10,000–50,000	Kholod et al. 2020
Sewage treatment	Up to 50,000 in sewers	Liu et al. 2015
Hydroelectric dams	Not available, but high dissolved concentrations	Kikuchi and Bingre do Amaral 2008; Reis et al. 2020
Dairy farms	Not available, but anaerobic digestors release methane	Wu et al. 2019; Centeno-Mora et al. 2020
Feedlots	Not available, but anaerobic digestors release methane	Wu et al. 2019; Centeno-Mora et al. 2020
Oil and gas wells	500–10,000 Up to 3000	Williams et al. 2019; Cho et al. 2020

ppm or greater (Table 1; Nisbet-Jones et al. 2022). There are many thousands of these emission sites in the United States alone, and examples of methane measurements suggest that a subset of such sites are candidates for methanotrophic methane removal (Table 1). It has been suggested that methane removal from the air at such sites using methanotroph-based technology is feasible (Yoon et al. 2009; Nisbet-Jones et al. 2022).

Methanotrophs have been used to remove methane from air using a biofilm-based bioreactor technology known as biofilters, but those systems are designed for methane in the 1%–5% (10,000–50,000 ppm) range (La et al. 2018) and do not contain methanotrophs adapted to the lower methane levels noted above (0.01%–0.1%). Biofilters consist of a solid support on which biofilms develop, often with countercurrent humidified gas streams and nutrient supply (La et al. 2018). A modeling study has predicted that a biofilter unit on the order of 100 m^3 treatment volume should have a capacity of ~6 tonnes methane/year at 500 ppm assuming 300 d operation/year (Yoon et al. 2009). More recent measurements (Ferdowsi et al. 2016) at 1000 ppm showed a capacity of 0.32 tonnes/m^3 treatment volume/year, which translates to 32 tonnes methane/year for a 100 m^3 unit. Assuming that methane removal is linear with concentration between 500 and 1000 ppm, we can predict half that amount (0.16 tonnes/

m^3/year; 16 tonnes/year/unit) at 500 ppm. The average of these studies for 500 ppm methane is ~0.1 tonnes/m^3/year, or for a unit of 100 m^3, 10 tonnes/year. This m^3-based capacity value (tonnes methane consumed/m^3 treatment volume/year) will be referred to in this paper as methane removal capacity and is related to the often-used term of elimination capacity (EC), which is g methane consumed/m^3/h, by a factor of 140.

Scale

As noted above, an impactful target for 2050 is 0.3 Gt cumulative methane removal, predicted to mitigate 0.07°C–0.22°C. Removing 15 Mt methane/year for 20 years would reach this goal. Assuming a 20-year deployment and an average of 10 tonnes/treatment unit/year capacity, such technology would require 1.5 million treatment units, and this may be more than the number of such emission sites worldwide. However, as noted above, current biofilter technology is not optimized for low methane levels, providing an opportunity for enhancement (see below). Increasing the methane removal capacity by 10-fold would drop the number required to 150,000 units worldwide, a goal that would be challenging but potentially feasible.

This analysis of concentration and scale is encouraging, suggesting that methane removal technology based on methanotrophs optimized for removing methane from air at 100–500 ppm or higher could theoretically be feasible at a globally meaningful scale.

METHANOTROPHIC BACTERIA

The organisms that consume methane are known as methanotrophs, and they comprise a group of bacteria and archaea with the ability to use methane as their sole carbon and energy source. The archaeal methanotrophs carry out reverse methanogenesis with electron acceptors of nitrate, sulfate, or metals (Evans et al. 2019), whereas the bacterial methanotrophs oxidize methane with O_2 via a methane monooxygenase (Fig. 1; Koo and Rosenzweig 2021). These aerobic methanotrophs are amenable to genetic and genomic manipulation (Puri et al. 2015), and metabolic engineering has been carried out for a variety of value-added products from methane (Chistoserdova and Kalyuzhnaya 2018; Nguyen and Lee 2021).

One important parameter for removing methane at 100–500 ppm concentrations is the whole-cell affinity for methane (apparent K_m, or $K_{m(app)}$), in which low $K_{m(app)}$ correlates with higher ability to consume low methane. Two unrelated methane monooxygenases are known, the soluble Fe enzyme (sMMO), which has a $K_{m(app)}$ of 30–50 μM, and the membrane-bound Cu enzyme (pMMO), which has a much lower

$K_{m(app)}$ of 1–7 μM (Kneif and Dunfield 2005; Tveit et al. 2019). As noted above, aerobic methanotrophs expressing the pMMO have been shown to consume methane at atmospheric concentrations (albeit at rates orders of magnitude lower than at 500 ppm), and an alternative whole-cell kinetic parameter, the specific affinity ($a°_s$) has been suggested as appropriate for such low methane concentrations (Kneif and Dunfield 2005; Tveit et al. 2019). Specific affinity is the slope of the Michaelis–Menten plot (reaction velocity vs. substrate concentration) in the linear range (at low concentrations), with high $a°_s$ denoting high methane consumption at low methane. In this case, high $a°_s$ correlates with the ability to consume low methane concentrations and is highest in methanotrophs expressing pMMO (Kneif and Dunfield 2005; Tveit et al. 2019). Reverse methanogenesis in the methanotrophic archaea requires much higher methane, with $K_{m(app)}$ values on the order of mM (Lu et al. 2019). For the purposes of removing methane at 100–500 ppm in air, aerobic methanotrophic bacteria using the pMMO are the targets of interest.

PROCESS SIMULATION AND SUSTAINABILITY ASSESSMENT

Early-stage research and development work into the technical feasibility of climate change miti-

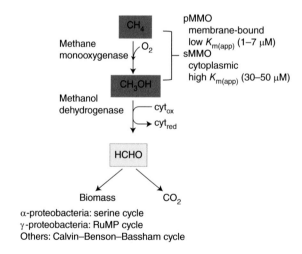

Figure 1. Methane metabolism in aerobic methanotrophs. (sMMO) soluble (cytoplasmic) methane monooxygenase, (pMMO) particulate (membrane-bound) methane monooxygenase, (RuMP) ribulosemonophosphate, (cyt) cytochrome.

gation technologies should include an assessment of the likely environmental impacts and economic costs, to ensure that the mitigation system has a high likelihood of improving the net greenhouse gas situation, and achieving cost parity with competing reduction options (Nisbet-Jones et al. 2022). With technology such as methanotroph-based methane removal, estimates of environmental and economic impact must come from detailed process models that characterize the full material and energy balances of the entire system, together with capital and operational cost factors for all unit operations (e.g., Handler and Shonnard 2018). These studies should include the economic and environmental value of a sustainable single-cell protein product (methane-derived bacterial biomass), which is used as a feed supplement for aquaculture and potentially for livestock. Methane-derived single-cell protein has an estimated value of $1600/tonne (El Abbadi et al. 2022). Such effort is needed as part of a holistic assessment of methanotroph-based methane removal technology.

OPPORTUNITIES

The looming climate crisis creates a sense of urgency to develop mitigation options quickly, so deployment can begin and continue for a sufficient amount of time to slow global warming over the next few decades. Although multiple strategies exist for methane mitigation (Jackson et al. 2021; Nisbet-Jones et al. 2022), none of them are ready to deploy at scale. An attractive feature of methanotroph-based methane removal is the history of commercial biofilter-based methane removal technology. This provides an existing platform of success on which to build enhanced systems, perhaps with a modified bioreactor configuration (see below). In addition, a methanotroph-based bioreactor strategy creates the possibility that a single technology could be deployed anywhere in the world. This situation creates optimism that methanotroph-based bioreactors for methane removal could become a reality in time to be a part of the solution to slow global warming.

Finally, the concern that enhancing methanotroph-based methane consumption in the

field in natural microbial communities will increase N_2O emissions adds impetus to create a closed system technology that can control both the microorganisms and their ancillary metabolism.

Costs

For methanotroph-based bioreactor technology to become feasible, it must be economically viable. A 2009 study (Yoon et al. 2009) assessed standard biofilter technology using methanotrophs to remove methane at 500 ppm based on literature values. They projected a financial cost for capital and operating of ~$500/tonne CO_2 equivalents (CO_2e) and CO_2 generation for operations equal to ~10% of the total CO_2e removed (assuming no heating). This analysis suggests improvements must be made in the methane consumed/treatment volume/year for such a biofilter approach to become economically feasible.

Conceptual Design

Figure 2 shows a conceptual design of a possible methane bioreactor technology. In this scheme, the air overlying or downwind of emission sites would be blown into a modular thin film–based bioreactor system containing a methane-utilizing bacterial community. The methane would be converted to CO_2 and biomass, and the biomass would be periodically harvested for use as a protein feedstock. Alternatively, methanotrophs could be engineered to excrete a value-added product such as a biofuel or other chemical feedstock (Chistoserdova and Kalyuzhnaya 2018; Nguyen and Lee 2021). If the blowers and bioreactor were low power and efficient and the water use minimized, such a technology could become both economically feasible and environmentally sound.

Potential Improvements

Both microbial and bioreactor improvements are possible for methanotroph-based bioreactor methane removal schemes. So far, published methane biofilter approaches have not utilized methanotrophs specifically chosen for their ability to consume methane at low levels. In the Lidstrom group, we have screened strains in our culture collection and discovered a subset with superior growth and methane oxidation rates at 100 and 500 ppm methane (He et al. 2023) compared to results in the literature (Kneif and Dunfield 2005). Further screens will undoubtedly identify even better strains for this purpose. It may also be possible to utilize both targeted and untargeted synthetic biology approaches to enhance methane consumption at 100–500 ppm methane by either altering the methane oxidation system (the pMMO), redirecting metabolic flux, or increasing metabolic efficiency.

For bioreactor improvements, approaches using thin film technology have been applied to microbial conversions (Padhi and Gokhale 2017; Ekins-Coward et al. 2019), and these have the advantage of high methane mass transfer and low water usage. Both are important parameters for a future deployable technology and have the potential to both increase methane removal capacity and decrease costs. Finally, advances that would result in low power consumption or utilization of solar panels and allow efficient harvesting of single-cell protein would also be expected to enhance

the performance and financial feasibility. Based on the analysis described above (Yoon et al. 2009), it would appear that a 10-fold methane removal capacity improvement would generate an economically feasible system achieving significant net CO_2e removal and targeting methane at 500 ppm, but as noted above, a complete assessment is necessary to test this prediction.

Risks

Because methanotroph-based biofilter technology is commercially viable at higher methane concentrations (La et al. 2018), the main risks involved in the approach described here are the inability to achieve a 10-fold increase in methane consumption and issues that might arise with scale-up of a modified bioreactor design.

NEEDS

For a methanotroph-based bioreactor methane removal technology to be achieved and commercialized, a set of steps must be taken. First, laboratory-scale studies are needed to carry out the en-

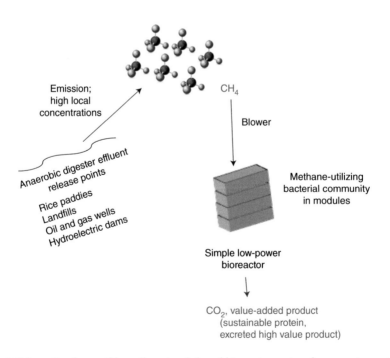

Figure 2. Schematic of a possible methanotroph-based bioreactor system for removing methane.

hancements noted above, and full-scale life cycle assessments and technoeconomic models need to be established. Several laboratories around the world have the needed expertise for the range of this required work, and potential strains and bioreactor technologies are already available, increasing the feasibility of such studies. Umbrella funding that brings multiple groups together under a shared technology goal would be necessary to achieve these goals in a 5-yr time frame. Based on this work, pilot systems would need to be developed, and once evaluated, a community-based program involving collaboration with those most affected by methane release could then result in deployment in the field, with further rounds of optimization. At that point, the technology should be ready to be commercialized and deployed at scale. Working together with communities strongly affected by methane release, such an approach has the potential to make a difference in global warming by 2050.

SUMMARY

Methane is a prime target for addressing the increasing urgency of climate change and global warming, through a combination of emissions reduction and methane removal from air. Although multiple approaches exist for methane removal, methanotroph-based bioreactor methane removal is potentially feasible in a relatively short time frame if funding becomes available to support the necessary work. In addition, it has the advantages of being modular, deployable at large numbers of sites globally, and potentially economically and environmentally viable. It will be important to develop technology in concert with communities affected by methane emissions to achieve mutually beneficial outcomes.

ACKNOWLEDGMENTS

This article has been made freely available online by generous financial support from the Bill and Melinda Gates Foundation. We particularly want to acknowledge the support and encouragement of Rodger Voorhies, president, Global Growth and Opportunity Fund.

REFERENCES

Abernethy S, O'Connor FM, Jones CD, Jackson RB. 2021. Methane removal and the proportional reductions in the surface temperature and ozone. *Philos Trans A Math Phys Eng Sci* **379**: 20210104. doi:10.1098/rsta.2021.0104

Börjesson G, Samuelsson J, Chanton J, Adolfsson R, Galle B, Svensson BH. 2009. A national landfill methane budget for Sweden based on field measurements, and an evaluation of IPCC models. *Tellus B* **61**: 424–435. doi:10.1111/j.1600-0889.2008.00409.x

Carmen RE, Vincent RK. 1998. Measurements of soil gas and atmospheric methane content on one active and two inactive landfills in Wood County, Ohio. *Environ Eng Geosci* **4**: 317–329. doi:10.2113/gseegeosci.IV.3.317

Centeno-Mora E, Fonseca PR, Andreão WL, Brandt EMF, de Souza CL, de Lemos Chernicharo C. 2020. Mitigation of diffuse CH4 and H2S emissions from the liquid phase of UASB-based sewage treatment plants: challenges, techniques, and perspectives. *Environ Sci Pollut Res Int* **27**: 35979–35992. doi:10.1007/s11356-020-08644-0

Chang J, Kim DD, Semrau JD, Lee JY, Heo H, Gu W, Yoon S. 2021. Enhancement of nitrous oxide emissions in soil microbial consortia via copper competition between proteobacterial methanotrophs and denitrifiers. *Appl Environ Microbiol* **87**: e0230120. doi:10.1128/AEM.02301-20

Chistoserdova L, Kalyuzhnaya MG. 2018. Current trends in methylotrophy. *Trends Microbiol* **26**: 703–714. doi:10.1016/j.tim.2018.01.011

Cho Y, Ulrich BA, Zimmerle DJ, Smits KM. 2020. Estimating natural gas emissions from underground pipelines using surface concentration measurements. *Environ Pollut* **267**: 115514. doi:10.1016/j.envpol.2020.115514

Dreyfus GB, Xu Y, Shindell DT, Zaelke D, Ramanathan V. 2022. Mitigating climate disruption in time: a self-consistent approach for avoiding both near-term and long-term global warming. *Proc Natl Acad Sci* **119**: e2123536119. doi:10.1073/pnas.2123536119

Ekins-Coward T, Boodhoo KVK, Velasquez-Orta S, Caldwell G, Wallace A, Barton R, Flickinger MC. 2019. A microalgae biocomposite-integrated spinning disk bioreactor (SDBR): toward a scalable engineering approach for bioprocess intensification in light-driven CO_2 absorption applications. *Ind Eng Chem Res* **58**: 5936–5949. doi:10.1021/acs.iecr.8b05487

El Abbadi SH, Sherwin ED, Brandt AR, Luby SP, Criddle CS. 2022. Displacing fishmeal with protein derived from stranded methane. *Nat Sustain* **5**: 47–56. doi:10.1038/s41893-021-00796-2

Evans PN, Boyd JA, Leu AO, Woodcroft BJ, Parks DH, Hugenholtz P, Tyson GW. 2019. An evolving view of methane metabolism in the Archaea. *Nat Rev Microbiol* **17**: 219–232. doi:10.1038/s41579-018-0136-7

Ferdowsi M, Veillette M, Ramirez AA, Jones JP, Heitz M. 2016. Performance evaluation of a methane biofilter under steady state, transient state and starvation conditions. *Water Air Soil Pollut* **227**: 168. doi:10.1007/s11270-016-2838-7

Handler RM, Shonnard DR. 2018. Environmental life cycle assessment of methane biocatalysis: key considerations and potential impacts. In *Methane biocatalysis: paving the way to sustainability*, pp. 253–270. Springer, Cham, Switzerland.

Harborth P, Fuß R, Münnich K, Flessa J, Fricke K. 2013. Spatial variability of nitrous oxide and methane emissions from an MBT landfill in operation: strong N$_2$O hotspots at the working face. *Waste Manag* **33**: 2099–2107. doi:10.1016/j.wasman.2013.01.028

He L, Groom JD, Wilson EH, Fernandez J, Konopka MC, Beck DAC, Lidstrom ME. 2023. A methanotrophic bacterium to enable methane removal for climate mitigation. *Proc Nat Acad Sci* **120**: e2310046120. doi:10.1073/pnas.2310046120

Intergovernmental Panel on Climate Change (IPCC). 2021. Climate change 2021: the physical science basis. In *Contribution of Working Group I to the Sixth Assessment Report of the Intergovernmental Panel on Climate Change* (ed. Masson-Delmotte V, Zhai P, Pirani A, Connors SL, Péan C, Berger S, Caud N, Chen Y, Goldfarb L, Gomis MI, et al.). Cambridge University Press, Cambridge (in press).

Jackson RB, Abernethy S, Canadell J, Cargnello M, Davis SJ, Féron S, Fuss S, Heyer AJ, Hong C, Jones CD, et al. 2021. Atmospheric methane removal: a research agenda. *Philos Trans A Math Phys Eng Sci* **379**: 20200454. doi:10.1098/rsta.2020.0454

Kang H, Lee J, Zhou X, Kim J, Yang Y. 2021. The effects of N enrichment on microbial cycling of non-CO$_2$ greenhouse gases in soils—a review and a meta-analysis. *Microb Ecol* **84**: 945–957. doi:10.1007/s00248-021-01911-8

Kholod N, Evans M, Pilcher RC, Roshchanka V, Ruiz F, Coté M, Collings R. 2020. Global methane emissions from coal mining to continue growing even with declining coal production. *J Clean Prod* **256**: 120489. doi:10.1016/j.jclepro.2020.120489

Kikuchi R, Bingre do Amaral P. 2008. Conceptual schematic for capture of biomethane released from hydroelectric power facilities. *Bioresour Technol* **99**: 5967–5971. doi:10.1016/j.biortech.2007.10.030

Kneif C, Dunfield P. 2005. Response and adaptation of different methanotrophic bacteria to low methane mixing ratios. *Environ Microbiol* **7**: 1307–1317. doi:10.1111/j.1462-2920.2005.00814.x

Koo CW, Rosenzweig AC. 2021. Biochemistry of aerobic methane oxidation. *Chem Soc Rev* **50**: 3424–3436. doi:10.1039/D0CS01291B

La H, Hettiaratchi JPA, Achari G, Dunfield PF. 2018. Biofiltration of methane. *Bioresour Technol* **268**: 759–772. doi:10.1016/j.biortech.2018.07.043

Limbri H, Gunawan C, Thomas T, Smith A, Scott J, Rosche B. 2014. Coal-packed methane biofilter for mitigation of green house gas emissions from coal mine ventilation air. *PLoS ONE* **9**: e94641. doi:10.1371/journal.pone.0094641

Liu Y, Ni BJ, Sharma KR, Yuan Z. 2015. Methane emission from sewers. *Sci Total Environ* **524**: 40–51. doi:10.1016/j.scitotenv.2015.04.029

Lu P, Liu T, Ni BJ, Guo J, Yuan Z, Hu S. 2019. Growth kinetics of *Candidatus* "Methanoperedens nitroreducens" enriched in a laboratory reactor. *Sci Total Environ* **659**: 442–450. doi:10.1016/j.scitotenv.2018.12.351

Nag M, Shimaoka T, Nakayama H, Komiya T, Xiaoli C. 2016. Field study of nitrous oxide production with in situ aeration in a closed landfill site. *J Air Waste Manag Assoc* **66**: 280–287. doi:10.1080/10962247.2015.1130664

Nguyen AD, Lee EY. 2021. Engineered methanotrophy: a sustainable solution for methane-based industrial biomanufacturing. *Trends Biotechnol* **39**: 381–396. doi:10.1016/j.tibtech.2020.07.007

Nisbet-Jones PBR, Fernandez JM, Fisher RE, France JL, Lowry D, Waltham DA, Woolley Maisch CA, Nisbet EG. 2022. Is the destruction or removal of atmospheric methane a worthwhile option? *Philos Trans A Math Phys Eng Sci* **380**: 20210108. doi:10.1098/rsta.2021.0108

Ocko IB, Sun T, Shindell D, Oppenheimer M, Hristov AN, Pacala SW, Mauzerall DL, Xu Y, Hamburg SP. 2021. Acting rapidly to deploy readily available methane mitigation measures by sector can immediately slow global warming. *Environ Res Lett* **16**: 054052. doi:10.1088/1748-9326/abe490

Padhi SK, Gokhale S. 2017. Treatment of gaseous volatile organic compounds using a rotating biological filter. *Bioresour Technol* **244**: 270–280. doi:10.1016/j.biortech.2017.07.112

Puri AW, Owen S, Chu F, Chavkin T, Beck DA, Kalyuzhnaya MG, Lidstrom ME. 2015. Genetic tools for the industrially promising methanotroph *Methylomicrobium buryatense*. *Appl Environ Microbiol* **81**: 1775–1781. doi:10.1128/AEM.03795-14

Reis PCJ, Ruiz-González C, Crevecoeur S, Soued C, Prairie YT. 2020. Rapid shifts in methanotrophic bacterial communities mitigate methane emissions from a tropical hydropower reservoir and its downstream river. *Sci Total Environ* **748**: 141374. doi:10.1016/j.scitotenv.2020.141374

Smith SJ, Chateau J, Dorheim K, Drouet L, Durand-Lasserve O, Frick O, Fujimori S, Hanaoka T, Harmsen M, Hilaire JK, et al. 2020. Impact of methane and black carbon mitigation on forcing and temperature: a multi-model scenario analysis. *Clim Change* **163**: 1427–1442. doi:10.1007/s10584-020-02794-3

Stein LY. 2020. The long-term relationship between microbial metabolism and greenhouse gases. *Trends Microbiol* **8**: 500–511. doi:10.1016/j.tim.2020.01.006

Sun T, Ocko IB, Sturcken E, Hamburg SP. 2021. Path to net zero is critical to climate outcome. *Sci Rep* **11**: 22173. doi:10.1038/s41598-021-01639-y

Tveit AT, Hestnes AG, Robinson SL, Schintlmeister A, Dedysh SA, Jehmlich N, von Bergen M, Herbold C, Wagner M, Richter A, et al. 2019. Widespread soil bacterium that oxidizes atmospheric methane. *Proc Natl Acad Sci* **116**: 8515–8524. doi:10.1073/pnas.1817812116

U.S. Department of State. 2021. Joint U.S.-EU statement on the global methane pledge. https://www.state.gov/joint-u-s-eu-statement-on-the-global-methane-pledge

Warszawski L, Kriegler E, Lenton TM, Gaffney O, Jacob D, Klingenfeld D, Koide RD, Máñez Costa M, Messner D, Nakicenovic N, et al. 2021. All options, not silver bullets, needed to limit global warming to 1.5°C: a scenario appraisal. *J Environ Res Lett* **16**: 064037. doi:10.1088/1748-9326/abfeec

Williams JP, Risk D, Marshall A, Nickerson N, Martell A, Creelman C, Grace M, Wach G. 2019. Methane emissions from abandoned coal and oil and gas developments in New Brunswick and Nova Scotia. *Environ Monit Assess* **191**: 479. doi:10.1007/s10661-019-7602-1

Cite this article as *Cold Spring Harb Perspect Biol* doi: 10.1101/cshperspect.a041671

Wu S, Li S, Zou Z, Hu T, Hu Z, Liu S, Zou J. 2019. High methane emissions largely attributed to ebullitive fluxes from a subtropical river draining a rice paddy watershed in China. *Environ Sci Technol* **53:** 3499–3507. doi:10 .1021/acs.est.8b05286

Ye M, Zheng W, Yin C, Fan X, Chen H, Gao Z, Zhao Y, Liang Y. 2022. The inhibitory efficacy of procyanidin on soil denitrification varies with N fertilizer type applied. *Sci Total Environ* **806:** 150588. doi:10.1016/j.scitotenv.2021 .150588

Yoon S, Carey JN, Semrau JD. 2009. Feasibility of atmospheric methane removal using methanotrophic biotrickling filters. *Appl Microbiol Biotechnol* **83:** 949–956. doi:10 .1007/s00253-009-1977-9

Agritech to Tame the Nitrogen Cycle

Lisa Y. Stein

Department of Biological Sciences, University of Alberta, Edmonton, Alberta T6G 2E9, Canada

Correspondence: Stein1@ualberta.ca; lisa.stein@ualberta.ca

While the Haber–Bosch process for N-fixation has enabled a steady food supply for half of humanity, substantial use of synthetic fertilizers has caused a radical unevenness in the global N-cycle. The resulting increases in nitrate production and greenhouse gas (GHG) emissions have contributed to eutrophication of both ground and surface waters, the growth of oxygen minimum zones in coastal regions, ozone depletion, and rising global temperatures. As stated by the Food and Agriculture Organization of the United Nations, agriculture releases ~9.3 Gt CO_2 equivalents per year, of which methane (CH_4) and nitrous oxide (N_2O) account for 5.3 Gt CO_2 equivalents. N-pollution and slowing the runaway N-cycle requires a combined effort to replace chemical fertilizers with biological alternatives, which after a 10-yr span of usage could eliminate a minimum of 30% of ag-related GHG emissions (~1.59 Gt), protect waterways from nitrate pollution, and protect soils from further deterioration. Agritech solutions include bringing biological fertilizers and biological nitrification inhibitors to the marketplace to reduce the microbial conversion of fertilizer nitrogen into GHGs and other toxic intermediates. Worldwide adoption of these plant-derived molecules will substantially elevate nitrogen use efficiency by crops while blocking the dominant source of N_2O to the atmosphere and simultaneously protecting the biological CH_4 sink. Additional agritech solutions to curtail N-pollution, soil erosion, and deterioration of freshwater supplies include soil-free aquaponics systems that utilize improved microbial inocula to enhance nitrogen use efficiency without GHG production. With adequate and timely investment and scale-up, microbe-based agritech solutions emphasizing N-cycling processes can dramatically reduce GHG emissions on short time lines.

CURRENT STATE OF THE GLOBAL N-CYCLE

Humanity is quickly approaching the moment where decisions on how we manage land use, distribution of planetary resources, and food supplies will determine our fate in the face of rapid climate change. It is clear that global inequality, wherein the wealthiest 10% of individuals are responsible for as much usage of resources and ecological damage as the bottom 80%, must be addressed if we are to avoid (or minimize) the more dire consequences to the human population and habitability of the planet (Rammelt et al. 2023). Without adjustments to our global food systems, negative environmental effects are expected to increase by 50%–90% from 2010 to 2050 due to further additions to the human population and a concomitant rise in global income (Springmann et al. 2018). The greatest environmental effect will be increased greenhouse gas (GHG) emissions followed by demands for arable land, water usage, and application of nitrogen and phosphorous fertilizers. Planetary boundaries that have already been crossed due to our current food systems include climate change, biosphere integrity, and N-cycle and phosphorous cycle thresholds (Rockstrom et al. 2009; Springmann et al. 2018). Additional boundaries that will be crossed soon include freshwater use, ocean acidification, land-use change, and destabilization of ecosystem functions that support human life.

Global food production is at the heart of our survival, which relies on delivering nutrients (e.g., nitrogen and phosphorous) to crops across diverse geographical regions. Prior to industrialized society, nitrogen delivery to plants was accomplished by microbial N_2-fixation and other natural processes. However, our relationship to the N-cycle changed when we discovered how to chemically fix atmospheric N_2 into reactive-N via the Haber–Bosch process (Galloway et al. 2014). Between 1850 and 1950, human-derived reactive-N (HdRn) was proportional to the population. From 1950 to 1980, the Green Revolution resulted in a marked uptick in HdRn, from about 12 to 30 $kg N yr^{-1} capita^{-1}$. This was also a period of rapid population growth, peaking at around a 2% growth rate in the 1960s (Ritchie et al. 2023). From 1980 to the present day, HdRn has stabilized to around 30 $kg N yr^{-1} capita^{-1}$, but due to a population of 8 billion and rising, this level of HdRN exceeds the planetary boundaries of sustainability, causing increased ecosystem degradation and unmitigated GHG production (Rockstrom et al. 2009; Galloway et al. 2014).

Some practical solutions to slow the crossing of planetary sustainability boundaries include shifting food choices toward plant-based diets, reducing food waste, and improving technology and management of food systems (Springmann et al. 2018). In addition, there is increasing interest in agritech solutions aimed toward direct reduction of GHG emissions while alleviating other environmental stresses (e.g., water and land demands). For instance, seaweed-based feed ingredients are under development to control CH_4 emissions from enteric fermentation in food animals (Vijn et al. 2020). Several companies are marketing microbial solutions to replace synthetic nitrogen and phosphorus fertilizers. Scaled-up soil-free systems that minimize water, chemical, land, and energy usage are emerging in urban centers. However, globalizing agritech requires overcoming substantial hurdles, including global social inequality, diversity of regional needs, geographical variations in ecosystems, and availability of investment and scale-up opportunities.

There are three main thresholds that must be considered for taming the global N-cycle: N-deposition rate, N-concentration in surface waters, and N-concentration in groundwaters (Schulte-Uebbing et al. 2022). However, extensive regional variation for these thresholds complicates mitigation strategies as Europe, China, India, and the United States experience N-excess, while N-deficiency is widespread in sub-Saharan Africa, Central and South America, and Southeast Asia. In N-deficient locations, none of the N-cycle thresholds have yet been crossed (Schulte-Uebbing et al. 2022). Biologically, there are also vast differences in how soils, plants, and microorganisms interact to manage nitrogen use efficiency depending on the health status of the soil. For instance, in nutrient-poor soils, plants have a strong dependence on fertilizer-N, whereas richer soils have ecologically complex networks of microbiota, plants, and minerals that support elaborate cycling and uptake of organic-N (Grandy et al. 2022). Thus, re-engineering of agroecosystems depends on regional differences in soil nutrition, management, and climate that can be partially addressed using a systems-biology understanding of plant-microbe-mineral interactions that impact regional nutrient cycles (Grandy et al. 2022).

MICROBIOLOGY OF THE N-CYCLE

Over the past several decades, researchers have made great strides in understanding the mechanisms, pathways, and ecology of the microbial N-cycle and the processes that generate N_2O (Stein 2019, 2020; Prosser et al. 2020). Chemolithotrophic ammonia oxidizers and denitrifiers are predominant and direct producers of N_2O in agroecosystems, although microbes encoding "clade II nitrous oxide reductase (NosZ)" enzymes can act as N_2O sinks as these microbes can use N_2O as their sole terminal electron acceptor (Fig. 1). Synthetic fertilizer usage has the strongest positive effect on the activity of ammonia-oxidizing bacteria (AOB), as these microorganisms tend to prefer inorganic over organic ammonia as their sole energy source (Hink et al. 2018a,b). Furthermore, AOB, unlike their archaeal counterparts (AOA), can reduce nitrite to N_2O enzymatically and are thus relatively efficient at producing N_2O as a terminal product; ca. up to 10% of ammonia-N can be transformed to N_2O-N by AOB given ample substrate and with low oxygen conditions (Stein 2019).

The product of ammonia oxidation by AOB and AOA is nitrite, which is further oxidized to nitrate by nitrite-oxidizing bacteria to complete the process of nitrification. Comammox bacteria oxidize ammonia to nitrate in one step, earning their title as "complete ammonia oxidizers" (Stein 2019). Nitrate is a primary pollutant and marker of N-saturation, contributing to eutrophication and contamination of surface and groundwaters that leads to oxygen minimum zone formation in coastal regions. Nitrate is also the first terminal electron acceptor for denitrification, which is the other major microbial source of N_2O (Fig. 1). As nitrification is both a direct and indirect source of N_2O, and a competitor for fixed nitrogen by plants, inhibitors of the nitrification process are essential for regions suffering from N-saturation effects (Norton and Ouyang 2019). While several options for nitrification inhibition have been explored, the use of biological nitrification inhibitors (BNIs) seems to hold promise for mitigating N_2O emissions

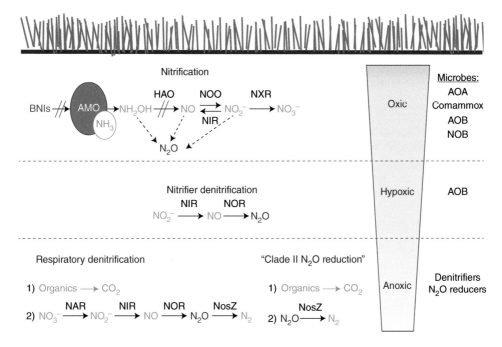

Figure 1. N-cycle processes and feedbacks in soil that contribute to N_2O. Biological nitrification inhibitors (BNIs) block the activity of AMO or HAO enzyme complexes (and can also block methane monooxygenases). Nitrification is the main nitrous oxide (N_2O)-producing process under oxic conditions. Nitrifier denitrification, performed by ammonia-oxidizing bacteria (AOB), produces N_2O under hypoxic conditions. Respiratory denitrification is the main N_2O-producing process under anoxic conditions and is driven by microbes that consume organic carbon. The process of "clade II N_2O reduction" removes N_2O to N_2 under anoxic conditions by microorganisms that consume organic carbon. (AMO) Ammonia monooxygenase, (HAO) hydroxylamine dehydrogenase, (NOO) nitric oxide oxidase, (NIR) nitrite reductase, (NXR) nitrite oxidoreductase, (NOR) nitric oxide reductase, (NAR) nitrate reductase, (NosZ) nitrous oxide reductase.

and causing the least harm to soil health (Nardi et al. 2020).

AGRITECH TO MITIGATE N_2O FROM AGRICULTURE

Biofertilizer is a rapidly growing, microbe-based, agritech venture, with profitable companies emerging worldwide. Biofertilizers make use of microorganisms that are natural members of the plant microbiome to augment seed germination and growth, resistance to pathogens, and overall yields and crop quality (Satish Kumar et al. 2022). The success of biofertilizers relies on the systematic evaluation of plant–microbe relationships and the composition of the rhizosphere microbiome. Beneficial microorganisms targeted for biofertilizers possess traits including the ability to degrade organic matter, enhancement of nutrient accessibility, such as phosphorous and nitrogen, production of phytohormones, and increased resistance to pathogen infection along with protection from other stresses. Once beneficial microorganisms are identified, they can be further engineered or used as natural products to apply directly to crops.

Some commercialized classes of biofertilizers include microbes that increase nitrogen fixation as free-living, associative, or symbiotic partners; microbes that solubilize phosphate, zinc, or potassium; sulfur-oxidizing microbes; and rhizobacteria that promote plant growth through secreted growth factors (Satish Kumar et al. 2022). Successful application of biofertilizer microbes into soils requires close attention to their quality, longevity, activity, purity, and associated carrier material to ensure their survival and adequate stimulation of crop growth and improved health (Satish Kumar et al. 2022). Although biofertilizers have been successfully produced and deployed as a general soil amendment for a variety of crop and soil types, the diversity of plant–microbe relationships across agroecosystems may need to be reconfigured from one farm to the next. In other words, a biofertilizer that works well for corn farmers in the United States might not work for rice farmers in the Philippines, thus requiring further research into the microbiome that naturally associates with target crops across diverse geographic regions.

A related crop amendment to biofertilizers is a group of compounds known as BNIs. Nitrification is the primary competitor with crops for fertilizer nitrogen, which has led to the application of synthetic nitrification inhibitors (SNIs), mainly nitrapyrin, dicyandiamide (DCD), and 3,4-dimethylpyrazole phosphate (DMPP) (Subbarao and Searchinger 2021). While these chemicals are effective in increasing nitrogen use efficiency and reducing N_2O emissions, they are also unstable and mainly affect the activity of AOB, and not AOA, which are often more dominant and active than AOB in soils (Leininger et al. 2006). Furthermore, SNIs can have deleterious effects on nontarget microbes in soils (Nardi et al. 2020), causing unintended alterations to soil functions. Contrary to SNIs, BNIs are naturally produced molecules exuded by plant roots into the rhizosphere to enable them to compete with nitrifiers for available ammonium (Nardi et al. 2020). BNIs target a variety of enzymes and structures in the microbial cell including the enzymes of nitrification (ammonia monooxygenase and hydroxylamine dehydrogenase), membranes, central metabolites (e.g., nitric oxide), and quorum-sensing molecules (Nardi et al. 2020).

Agritech ventures aim to identify BNIs produced naturally by target crops that inhibit both AOB and AOA. For application, BNI production in situ by crops can be enhanced using genetic engineering tools, or by direct amendment of soils with industrially produced BNIs. Key research required to bring BNIs to the marketplace includes (1) ensuring their efficacy, either alone or in cocktails, in both improving nitrogen use efficiency and in preventing N_2O production, (2) ensuring that they do not impair other soil services by affecting nontarget microbiota, and (3) ensuring that they do not target methane oxidizers that act as the sole biological CH_4 sink (Stein and Klotz 2011). To this last point, BNIs may dramatically reduce N_2O emissions while stimulating CH_4 emissions due to inhibition of the homologous enzyme complexes, ammonia and methane monooxygenase (Klotz and Norton 1998).

Soil-free cropping systems are another agritech solution that addresses multiple sustainability boundaries, including nitrogen and phosphorous cycles and freshwater use, among others. The most promising emergent soil-free technology is aquaponics, which integrates fish with crop production (Goddek et al. 2019). In recirculating aquaponics systems, fish waste is converted by microorganisms to vital plant nutrients, mainly nitrate, and the plants regenerate clean water for the fish. These enclosed systems can operate at low energy, low water usage, and occupy minimal land space while producing substantial amounts of leafy green crops and fish protein (Goddek et al. 2019). Although aquaponics technology is immature, it is clear that further developments to improve nitrogen use efficiency and microbiome function to avoid N_2O production and pathogen infection is needed, along with consideration of socioeconomic aspects for integrating soil-free farming in the context of regional and geographic diversity (Junge et al. 2017). There is also a need to better explore plant-microbe interactions and improvements to microbial inocula to expand the variety of crops that can be produced at high efficiency while improving nitrogen use efficiency and reducing N_2O emissions.

SYNERGIES BETWEEN N-CYCLE AGRITECH WITH OTHER APPLICATIONS

Many other agritech solutions, including crop engineering for enhanced photosynthesis, synthetic biology solutions to improve rates of carbon fixation, reduction of stomata in food crops, and increasing carbon-efficient tree plantations, are vital to slow the crossing of planetary sustainability boundaries in food production systems. However, keen attention to the microbial N-cycle in all agritech has the capacity to mitigate N_2O emissions and improve overall nitrogen use efficiency by crops, thus bringing the N-cycle back toward sustainability. The N-cycle is intimately connected to the carbon cycle, such that impacts or interference with one cycle will necessarily impact the other. For instance, the microbial processes responsible for N_2O and CH_4 emissions are both sensitive to redox and nutrient balance and are also heavily influenced by the availability of trace metals like copper and iron (Stein 2020). As an example, methanotrophic bacteria produce the copper-chelating chalkophore, metha-

nobactin, that prevents access to copper for the expression of nitrous oxide reductase enzymes (NosZ) by denitrifiers (Chang et al. 2021). Thus, when copper is at low availability, CH_4 oxidation will proceed at the expense of N_2O reduction, leading to higher N_2O emissions with lower CH_4 emissions. Similarly, some methane oxidizers use nitrate as a terminal electron acceptor, leading to N_2O as a final product (Kits et al. 2015); the result is a reduction in CH_4 emission with a proportional increase in N_2O emission when nitrate is readily available and low redox conditions promote its usage. Therefore, technologies aimed at reducing CH_4 emissions must consider how nutrient amendments, like copper and nitrate, effect microbial processes leading to N_2O emissions, and vice versa.

Microbial agritech ventures are undergoing a revolution that aims to bring the N-cycle back toward sustainability. The success of the technologies discussed above rely heavily on investment and scale-up, specific regional and geographic needs, and attention to global socioeconomic inequalities. Research focused on systems biology, the integration of plant-microbe-mineral interactions, and microbiome function is necessary to achieve widespread development, optimization, and deployment of microbial agritech. Specific attention to the microbial N-cycle and its interrelationships to other nutrient cycles, including carbon, is essential for disrupting our current food production systems and replacing them with highly efficient, biology-based systems that work within the boundaries of planetary sustainability.

ACKNOWLEDGMENTS

This article has been made freely available online by generous financial support from the Bill and Melinda Gates Foundation. We particularly want to acknowledge the support and encouragement of Rodger Voorhies, president, Global Growth and Opportunity Fund.

REFERENCES

Chang J, Kim DD, Semrau JD, Lee JY, Heo H, Gu WY, Yoon S. 2021. Enhancement of nitrous oxide emissions in soil microbial consortia via copper competition between proteobacterial methanotrophs and denitrifiers. *Appl Environ Microbiol* **87**: e0230120. doi:10.1128/AEM.02301-20

Galloway JN, Winiwarter W, Leip A, Leach AM, Bleeker A, Erisman JW. 2014. Nitrogen footprints: past, present and future. *Environ Res Lett* **9**: 115003. doi:10.1088/1748-9326/9/11/115003

Goddek S, Joyce A, Kotzen B, Burnell G. 2019. Aquaponics food production systems: combined aquaculture and hydroponic production technologies for the future. In *Aquaponics food production systems*, p. 619. Springer, New York. doi:10.1007/978-3-030-15943-6

Grandy AS, Daly AB, Bowles TM, Gaudin ACM, Jilling A, Leptin A, McDaniel MD, Wade J, Waterhouse H. 2022. The nitrogen gap in soil health concepts and fertility measurements. *Soil Biol Biochem* **175**: 108856. doi:10.1016/j.soilbio.2022.108856

Hink L, Nicol GW, Prosser JI. 2018a. Archaea produce lower yields of N$_2$O than bacteria during aerobic ammonia oxidation in soil. *Environ Microbiol* **19**: 4829–4837. doi:10.1111/1462-2920.13282

Hink L, Gubry-Rangin C, Nicol GW, Prosser JI. 2018b. The consequences of niche and physiological differentiation of archaeal and bacterial ammonia oxidisers for nitrous oxide emissions. *ISME J* **12**: 1084–1093. doi:10.1038/s41396-017-0025-5

Junge R, König B, Villarroel M, Komives T, Jijakli MH. 2017. Strategic points in aquaponics. *Water (Basel)* **9**: 182. doi:10.3390/w9030182

Kits KD, Klotz MG, Stein LY. 2015. Methane oxidation coupled to nitrate reduction under hypoxia by the Gammaproteobacterium *Methylomonas denitrificans*, sp. nov. type strain FJG1. *Environ Microbiol* **17**: 3219–3232. doi:10.1111/1462-2920.12772

Klotz MG, Norton JM. 1998. Multiple copies of ammonia monooxygenase (*amo*) operons have evolved under biased AT/GC mutational pressure in ammonia-oxidizing autotrophic bacteria. *FEMS Microbiol Lett* **168**: 303–311. doi:10.1111/j.1574-6968.1998.tb13288.x

Leininger S, Urich T, Schloter M, Schwark L, Qi L, Nicol GW, Prosser JI, Schuster SC, Schleper. 2006. Archaea predominate among ammonia-oxidizing prokayotes in soils. *Nature* **442**: 806–809. doi:10.1038/nature04983

Nardi P, Laanbroek HJ, Nicol GW, Renella G, Cardinale M, Pietramellara G, Weckwerth W, Trinchera A, Ghatak A, Nanipieri P. 2020. Biological nitrification inhibition in the rhizosphere: determining interactions and impact on microbially mediated processes and potential applications. *FEMS Microbiol Rev* **44**: 874–908. doi:10.1093/femsre/fuaa037

Norton J, Ouyang Y. 2019. Controls and adaptive management of nitrification in agricultural soils. *Front Microbiol* **10**: 1931. doi:10.3389/fmicb.2019.01931

Prosser JI, Hink L, Gubry-Rangin C, Nicol GW. 2020. Nitrous oxide production by ammonia oxidizers: physiological diversity, niche differentiation and potential mitigation strategies. *Glob Change Biol* **26**: 103–118. doi:10.1111/gcb.14877

Rammelt CF, Gupta J, Liverman D, Scholtens J, Ciobanu D, Abrams JF, Bai X, Gifford L, Gordon C, Hurlbert M, et al. 2023. Impacts of meeting minimum access on critical earth systems amidst the Great Inequality. *Nat Sustain* **6**: 212–221. doi:10.1038/s41893-022-00995-5

Ritchie H, Rodés-Guirao L, Mathieu E, Gerber M, Ortiz-Ospina E, Hasell J, Roser M. 2023. Population growth. *Our World in Data.* https://ourworldindata.org/world-population-growth

Rockström J, Steffen W, Noone K, Persson A, Chapin FS III, Lambin EF, Lenton TM, Scheffer M, Folke C, Schellnhuber HJ, et al. 2009. A safe operating space for humanity. *Nature* **461**: 472–475. doi:10.1038/461472a

Satish Kumar D, Sindhu SS, Kumar R. 2022. Biofertilizers: an ecofriendly technology for nutrient recycling and environmental sustainability. *Curr Res Microb Sci* **3**: 100094.

Schulte-Uebbing LF, Beusen AHW, Bouwman AF, de Vries W. 2022. From planetary to regional boundaries for agricultural nitrogen pollution. *Nature* **610**: 507–512. doi:10.1038/s41586-022-05158-2

Springmann M, Clark M, Mason-D'Croz D, Wiebe K, Bodirsky BL, Lassaletta L, de Vries W, Vermeulen SJ, Herrero M, Carlson KM, et al. 2018. Options for keeping the food system within environmental limits. *Nature* **562**: 519–525. doi:10.1038/s41586-018-0594-0

Stein LY. 2019. Insights into the physiology of ammonia-oxidizing microorganisms. *Curr Opin Chem Biol* **49**: 9–15. doi:10.1016/j.cbpa.2018.09.003

Stein LY. 2020. The long-term relationship between microbial metabolism and greenhouse gases. *Trends Microbiol* **28**: 500–511. doi:10.1016/j.tim.2020.01.006

Stein LY, Klotz MG. 2011. Nitrifying and denitrifying pathways of methanotrophic bacteria. *Biochem Soc Trans* **39**: 1826–1831. doi:10.1042/BST20110712

Subbarao GV, Searchinger TD. 2021. A "more ammonium solution" to mitigate nitrogen pollution and boost crop yields. *Proc Natl Acad Sci* **118**: e2107576118. doi:10.1073/pnas.2107576118

Vijn S, Compart DP, Dutta N, Foukis A, Hess M, Hristov AN, Kalscheur KF, Kebreab E, Nuzhdin SV, Price NN, et al. 2020. Key considerations for the use of seaweed to reduce enteric methane emissions from cattle. *Front Vet Sci* **7**: 597430. doi:10.3389/fvets.2020.597430

Cite this article as *Cold Spring Harb Perspect Biol* doi: 10.1101/cshperspect.a041668

Microbial Catalysis for CO_2 Sequestration: A Geobiological Approach

Martin Van Den Berghe,[1,7] Nathan G. Walworth,[2,3,4,7] Neil C. Dalvie,[5] Chris L. Dupont,[4,6] Michael Springer,[5] M. Grace Andrews,[2] Stephen J. Romaniello,[2] David A. Hutchins,[3] Francesc Montserrat,[2] Pamela A. Silver,[5] and Kenneth H. Nealson[2,3]

[1]Cytochrome Technologies Inc., St. John's, Newfoundland and Labrador A1C 5S7, Canada

[2]Vesta, San Francisco, California 94114, USA

[3]University of Southern California, Los Angeles, California 90007, USA

[4]Department of Environment and Sustainability, J. Craig Venter Institute, La Jolla, California 92037, USA

[5]Department of Systems Biology, Harvard Medical School, Boston, Massachusetts 02115, USA

[6]Department of Human Biology and Genomic Medicine, J. Craig Venter Institute, La Jolla, California 92037, USA

Correspondence: nate@vesta.earth

One of the greatest threats facing the planet is the continued increase in excess greenhouse gasses, with CO_2 being the primary driver due to its rapid increase in only a century. Excess CO_2 is exacerbating known climate tipping points that will have cascading local and global effects including loss of biodiversity, global warming, and climate migration. However, global reduction of CO_2 emissions is not enough. Carbon dioxide removal (CDR) will also be needed to avoid the catastrophic effects of global warming. Although the drawdown and storage of CO_2 occur naturally via the coupling of the silicate and carbonate cycles, they operate over geological timescales (thousands of years). Here, we suggest that microbes can be used to accelerate this process, perhaps by orders of magnitude, while simultaneously producing potentially valuable by-products. This could provide both a sustainable pathway for global drawdown of CO_2 and an environmentally benign biosynthesis of materials. We discuss several different approaches, all of which involve enhancing the rate of silicate weathering. We use the silicate mineral olivine as a case study because of its favorable weathering properties, global abundance, and growing interest in CDR applications. Extensive research is needed to determine both the upper limit of the rate of silicate dissolution and its potential to economically scale to draw down significant amounts (Mt/Gt) of CO_2. Other industrial processes have successfully cultivated microbial consortia to provide valuable services at scale (e.g., wastewater treatment, anaerobic digestion, fermentation), and we argue that similar economies of scale could be achieved from this research.

---

[7]These authors contributed equally to this work.

Cite this article as *Cold Spring Harb Perspect Biol* doi: 10.1101/cshperspect.a041673

GLOBAL CLIMATE CHANGE

Human-driven global climate change is increasingly apparent, with global impacts accelerating during the last decade (Dreyfus et al. 2022). Over geological time, the global carbon cycle has regulated Earth's climate via silicate chemical weathering and carbonate formation (Fig. 1; Renforth and Henderson 2017). Weathering of silicate minerals causes uptake and storage of excess atmospheric CO_2 in various forms, including soluble dissolved inorganic carbon (DIC) and insoluble minerals like calcite. Today, the release of anthropogenic greenhouse gas (GHG) emissions derives from diverse human activities such as the burning of fossil fuels, deforestation, land use, cement production, etc. The rate of release of GHGs like carbon dioxide (CO_2) and methane are currently far too high to be buffered by both short-term (e.g., biological processes) and long-term (e.g., mineral weathering) carbon cycles, thereby pushing the Earth to multiple climate tipping points (Lenton et al. 2019).

Mitigating global climate change requires the cessation of the use of fossil fuels, which will slow the accumulation of GHGs in the atmosphere.

Despite some incremental progress, this has proven to be a difficult social and political task (Arora and Mishra 2021). Further, cessation of fossil fuel use is not enough to curb global warming. It is estimated that the concentration of CO_2 in the atmosphere will not decline for centuries (Cawley 2011), thereby leading to continuous, severe impacts even if emissions cease. Additionally, the transition to a green economy will require burning fossil fuels for some time into the future. Thus, major carbon dioxide removal (CDR) and storage (sequestration) efforts are necessary to limit global warming to 1.5°C–2°C more than preindustrial levels (National Academies of Sciences, Engineering, and Medicine 2022). To avoid the worst impacts of climate change, CDR methods must extract >1000 gigatonnes of CO_2 ($GtCO_2$) by 2100 (Campbell et al. 2022). To this end, a number of solutions have been suggested and modeled to be economically and operationally sustainable (Dreyfus et al. 2022).

GEOCHEMICAL CDR AND ENHANCED WEATHERING

In parallel to other CDR methods, we suggest that a global effort to accelerate the removal of

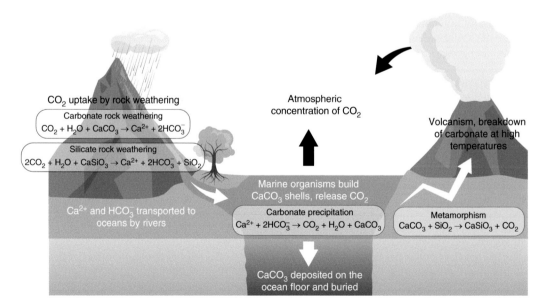

Figure 1. The carbonate–silicate cycle. CO_2 is sequestered in the form of HCO_3^- by rock weathering, transported to oceans, and eventually precipitated as $CaCO_3$ by marine organisms.

Cite this article as *Cold Spring Harb Perspect Biol* doi: 10.1101/cshperspect.a041673

CO_2 via enhanced silicate weathering should be initiated. Silicate weathering is widely recognized as one of the primary natural processes that regulates long-term atmospheric CO_2 concentrations (Renforth and Henderson 2017). Increasing the rate of weathering of silicate minerals (in this case, focusing on olivine) would accelerate naturally occurring geologic drawdown and storage of atmospheric CO_2. Collectively, terrestrial and marine silicate weathering are estimated to have the potential to sequester tens of billions of tonnes of CO_2/year (Renforth 2019; Campbell et al. 2022; National Academies of Sciences, Engineering and Medicine 2022). Both terrestrial and marine applications can have the potential to be more permanent and scalable carbon removal solutions than other widespread carbon removal approaches (e.g., afforestation) (Campbell et al. 2022), particularly because of the long residence time of DIC in the oceans (>10,000 yr) (Middelburg et al. 2020). Accordingly, ocean DIC is the largest DIC pool on Earth. In its 2021–2022 ocean-based CDR report, The National Academies of Sciences, Engineering and Medicine (2022) called for a rapid increase in research on ocean alkalinity enhancement (OAE) methods, inclusive of silicate-based CO_2 removal to help sequester >1000 $GtCO_2$ by 2100.

CO_2 drawdown from silicate weathering is principally controlled by the generation of alkalinity. Anions released into solution from mineral dissolution act as a proton sink (Eq. 1), neutralizing acidity present in an environment. In most natural aqueous environments, carbonic acid, resulting from the presence of atmospheric CO_2, is the dominant form of acidity. Hence, the generation of alkalinity in the marine environment induces the drawdown of atmospheric CO_2 following air–sea equilibration (He and Tyka 2022) to form DIC, principally as bicarbonate and carbonate ions:

$$Mg_{2-x}Fe_xSiO_4 + 4CO_2 + 4H_2O \rightarrow$$
$$2 - xMg^{2+} + xFe^{2+} + 4HCO_3^- + H_4SiO_4.$$
$$(1)$$

Bicarbonate ions (HCO_3^-) in seawater can have residence times in excess of 10,000 yr (Mid-

delburg et al. 2020), therefore qualifying as a highly permanent method for CO_2 removal and storage. Subsequently, marine calcification processes can convert DIC and alkalinity into solid carbonate species, such as calcite, aragonite, and vaterite. Although these secondary carbonate reactions decrease CO_2 removal efficiencies, they represent another permanent form of CO_2 sequestration as solid carbonate. Thus, both seawater DIC and carbonate precipitates ultimately offer an effective form of permanent storage.

Although various highly reactive hydroxide salts such as brucite also generate alkalinity and capture carbon effectively, they are not widely available in large enough quantities to contribute to gigatonne-scale CDR (Power et al. 2013). The most useful silicates are those that can both effectively capture the most protons per mole of silicate dissolution and dissolve the fastest. These include mafic–ultramafic neso-, soro-, and inosilicates minerals such as olivine and pyroxenes, as well as some hydrated clays. Of these, the minerals that have some of the highest dissolution rates are olivine, serpentine, and wollastonite (Hartmann et al. 2013; Bullock et al. 2022). They are globally abundant and occur in areas easily accessible to humans. Combined, the CDR potential of their known reserves is estimated to be in the tens of thousands of $GtCO_2$ (Lackner 2002; Power et al. 2013), orders of magnitude greater than what is needed to permanently sequester anthropogenic CO_2 emissions. The questions surrounding methods of carbon capture via ultramafic minerals therefore relate to the scientific, engineering, political, and economic logistics of implementation at large scales, which we explore below.

Olivine

Olivine is considered to be one of the most favorable silicate minerals for CDR, as it weathers quickly under Earth surface conditions (Rimstidt et al. 2012; Hartmann et al. 2013). This reactivity, along with its structure of silicate tetrahedra isolated by only two cations (Mg^{2+}, Fe^{2+}), makes olivine dissolution both relatively simple and highly energetically favorable. Additionally, olivine dissolution generates more alkalinity (principally HCO_3^-) than other common silicate minerals, with up to 4 mol of CO_2 sequestered per mol of olivine, equivalent to ~1 tonne CO_2 per tonne of olivine dissolved (Eq. 1). Hence, olivine ($Mg_{2-x}Fe_xSiO_4$) has been proposed as a scalable mineral for CDR applications, as it is a globally abundant, naturally occurring ultramafic silicate mineral (Beerling et al. 2021).

Microbial Catalysis of Olivine Dissolution

The rate of alkalinity generation during enhanced weathering is, in part, determined by the rate of silicate mineral dissolution (Rimstidt et al. 2012; Oelkers et al. 2018). Many studies of olivine (or other silicate minerals) conducted under laboratory conditions clearly indicate that weathering rates can be increased by grinding minerals to small particle sizes to increase the surface-to-volume ratio (Rosso and Rimstidt 2000). However, accelerated dissolution rates are difficult to sustain in abiotic closed laboratory systems, as demonstrated by us (Fig. 2) and others (Oelkers et al. 2015, 2018); and for reasons not yet clear, rates slowed significantly over time. Prior research showed that a nanometer-thick layer of ferric oxyhydroxide can form around the olivine as it weathers, acting as an impediment to further weathering (Schott and Berner 1983; Oelkers et al. 2018). Utilizing catalysts or implementing open-system applications like coastal enhanced weathering (Meysman and Montserrat 2017; Montserrat et al. 2017) may be able to accelerate the dissolution of olivine and other silicate minerals, thereby improving the feasibility and potential scale of CDR efforts based on enhanced weathering.

Despite the tremendous potential of silicate weathering for CDR, natural silicate dissolution needs to be greatly accelerated to limit global warming to 2°C more than preindustrial levels by the end of this century. Current estimates of natural, abiotic olivine dissolution place the half-life of a sand-sized olivine grain in the range of decades, nearing an asymptotic 100% dissolution of more than two centuries (Montserrat et al. 2017). Alternatively, using microbes as catalysts has the potential to reduce the mineral half-life to a few years, nearing an asymptotic 100% dissolution of ~50 yr (Berghe et al. 2021). A handful of studies tested the biological enhancement of weathering, and preliminary evidence demonstrates both fungi and bacteria can accelerate silicate dissolution via several mechanisms (Campbell et al. 2022 and references therein). Our experiments (Fig. 3) and those of others (e.g., Torres et al. 2014, 2019; Lunstrom et al. 2023) point to significant catalysis by Fe-binding organic ligands known as siderophores produced by several different bacterial groups. Whereas us-

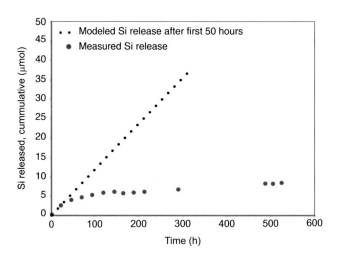

Figure 2. Modeled abiotic olivine dissolution rates in the first 50 h can be more than an order of magnitude greater than empirical dissolution rates over longer timescales, which become asymptotic.

 Cite this article as *Cold Spring Harb Perspect Biol* doi: 10.1101/cshperspect.a041673

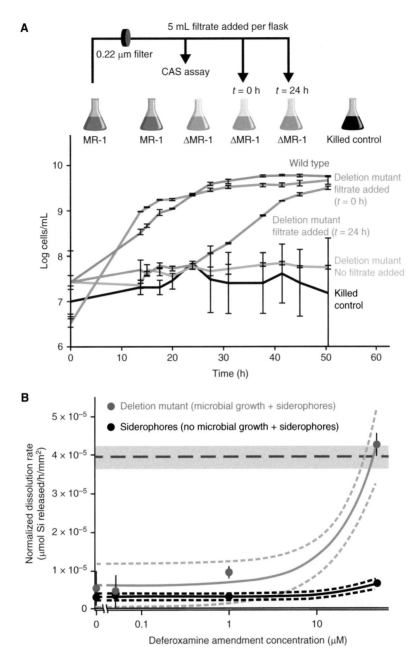

Figure 3. Plots showing *Shewanella oneidensis* growth dynamics and olivine dissolution rates across various experimental treatments. (*A*) Plot showing *S. oneidensis* growth from siderophore-based iron acquisition of olivine minerals, with the siderophore synthesis deletion mutant growing only when given exogenous siderophores. (Plot adapted with permission from Van Den Berghe et al. 2021, © 2021 John Wiley & Sons Ltd.) (*B*) Plot comparing normalized olivine dissolution rates as a function of siderophore concentrations and microbial activity. The red dashed line denotes wild-type *S. oneidensis* dissolution rates (native siderophore concentration/production unknown). The light gray line denotes olivine dissolution rates of microbial growth of the *S. oneidensis* deletion mutant as a function of added concentrations of siderophores. The black line denotes olivine dissolution rates as a function of increasing siderophore concentrations with no microbial growth. Dashed line with color background represents 95% confidence intervals on the regression lines.

ing microbes as catalysts offers a promising path forward, mineral dissolution rates must be characterized and measured against microbial carbon cycling and net carbon fluxes to fully characterize CDR potential. Here, we suggest a path for a sustainable CO_2 sequestration program using siderophores as a case study for one potentially promising mechanism to facilitate microbial catalysis of silicate weathering.

Siderophores are organic ligands with extremely high binding affinity for Fe^{3+} and intermediate affinities for Fe^{2+}, Mn^{3+}, and many other transition metals (Hider and Kong 2010). There is a wide variety of structures produced by many different bacteria and fungi as a component of their high-affinity Fe uptake systems, particularly in aerobic marine environments, where the solubility of iron is calculated to be 10^{-16} M.

Given that the formation constants for many siderophores are 10^{30} and higher, it is a microbial mechanism that is very prevalent in the ocean, in soils, and in at least a significant fraction of the microbes in the human (Manck et al. 2022). In the ocean, iron availability has a clear and direct impact on primary productivity (Hutchins and Boyd 2016), despite being absolutely insoluble in oxic environments ($K_{sp} = 10^{-37}$–10^{-44}; Guerinot and Yi 1994). This points to evolutionary mechanisms of microbial nutrient acquisition from insoluble mineral phases (Manck et al. 2022). In fact, this is what led to previous experimental efforts to apply ocean iron fertilization as a means for CDR (Williamson et al. 2012), whereby particulate Fe-containing materials can act as sources of scarce micronutrients, promoting primary productivity and microbial growth.

Experiments by us and others have shown that the addition of siderophores to olivine particles enhances the rate of the reaction (Torres et al. 2014, 2019; Berghe et al. 2021). Although this work was encouraging, the concentrations of siderophores required to sustain the reaction were high and likely prohibitively expensive for any large-scale CDR operations. Subsequent studies (Berghe et al. 2021; Lunstrum et al. 2023), however, showed that microbes that produce siderophores can obtain nutrients from olivine, resulting in sustained growth as a function of olivine weathering (Fig. 3A). Specifically, *She-*

wanella oneidensis obtained nutrients for growth from olivine particles only during the production of its siderophore, putrebactin (Ledyard and Butler 1997). This live system resulted in significantly higher rates of olivine dissolution relative to those previously seen with the addition of exogenous siderophores alone (Fig. 3B).

FUTURE DIRECTIONS

Prior experiments demonstrated increases in olivine dissolution rates by more than an order of magnitude via siderophore synthesis, but several key biological and engineering questions remain before CDR by microbially mediated mineral dissolution can be implemented at scale. Below, we outline important areas of research necessary to understand the full potential of microbially enhanced weathering.

The Biological Underpinnings of Mineral Dissolution

The precise mechanisms involved in the microbial acquisition of particulate-bound nutrients are not yet fully understood. Siderophores can play a crucial role in transforming mineral-phase iron (esp. oxidized iron) into a bioavailable form, which can remove a thin (nanometer-thick) layer of oxidized iron deposited along the surface of olivine grains. This process enabled microbial exponential growth for siderophore-producing bacteria, whereas bacteria incapable of producing siderophores proved to be unable to grow in the same conditions (Berghe et al. 2021). This accelerated dissolution seemed to be most pronounced during microbial exponential growth, which would be the period of maximal demand for new iron entering the biological pool.

Siderophore activity can be augmented by other, complementary microbial processes related to the formation of biofilms along mineral surfaces. Bacteria can form thick, redox-active biofilms along mineral surfaces (Fig. 4) that can support abundant and dense growth (Hall-Stoodley et al. 2004). Although some studies have suggested that biofilms can inhibit mineral dissolution (Oelkers et al. 2015), the success of these biofilms in sustaining growth in environ-

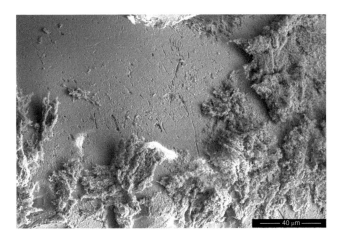

Figure 4. Scanning electron microscope (SEM) secondary electron imagery showing *Pseudomonas aeruginosa* biofilm growth atop olivine grains in an iron-deplete medium. Accelerated weathering features are clearly visible on the mineral surface (middle of the image) after only 48 h. Scale bar, 40 μm.

ments where dissolved nutrients are limiting suggests active mechanisms of nutrient acquisition directly from mineral phases. Biofilms can host a wide range of metabolic activities. If this activity can be channeled toward mineral-phase nutrient acquisition, the sheer density and intimate proximity of biofilms to mineral surfaces can have a profound effect on mineral dissolution rates, as observed in prior studies (Berghe et al. 2021; Campbell et al. 2022). In addition to siderophore activity, respiration can significantly lower the pH along the mineral surface, further enhancing mineral dissolution rates. Bacteria can also produce significant amounts of organic acids within the confines of biofilms (e.g., lactate, pyruvate, citrate; Rogers et al. 2013), which can have significant effects on localized pH and are often excellent cation sorbents that can further help promote mineral dissolution (Pokrovsky et al. 2009).

Here, we propose that a genetic and physiological understanding of mineral dissolution mechanisms would critically inform the utilization and application of microbially enhanced dissolution reactions. Identification of genetic synthesis pathways and their regulators would inform engineering efforts to improve dissolution. Particularly, a mechanistic understanding would drive the selection of promising organisms for large-scale dissolution systems, the application of genetic engineering to maximize dissolution, and the design of culture environments and reactor systems.

The Upper Limit of Microbially Mediated Dissolution and Biosynthetic Production

Sustaining microbially accelerated mineral dissolution requires the continuous maintenance of an environment that favors both a specific metabolic activity from a microbiome and the kinetics and thermodynamics of mineral dissolution and carbon drawdown. Scientific and technological advancements have transformed these challenges into engineering questions rather than scientific barriers. For example, maintaining an environment in a steady geochemical state to promote a specific reaction is done routinely (and at scale)—for example, in water treatment plants and breweries. Similarly with microbes, impressive advances in the field of synthetic biology now allow us to more easily genetically engineer non-model strains, although the future goals are to control entire microbiomes (Brophy et al. 2018; Lawson et al. 2019; Lee et al. 2020).

Microorganisms may also add separate economic value beyond the generation of alkalinity from mineral dissolution. These can include the biosynthesis of lipids, vitamins, and other biomolecules that have societal and economic value in industries such as pharmaceuticals or agriculture and aquaculture (Włodarczyk et al. 2020). Alternatively, certain substrates introduced into the reactors can promote the growth of a specific strain or metabolism, such as certain degradable plastics (e.g., polyhydroxyalkanoates [PHAs] and polyhydroxybutyrates [PHBs]) as a carbon source for biofilm-forming heterotrophs (Suzuki et al. 2021). Not only would such specific substrates select for specific, desired microbial activity, they can offer a sustainable means to manage plastic wastes while replacing the costs of adding a carbon source to the reactor. Last, the dissolution of large quantities (i.e., gigatonnes) of minerals will release significant amounts of trace metals associated with ultramafic materials, most notably nickel, cobalt, and chromium. These metals are particularly valuable, as they play a critical role in the transition away from fossil fuels and toward an electricity-based economy (Diallo et al. 2015).

We propose that microorganisms can be genetically engineered for large-scale mineral dissolution processes, specifically for production of siderophores, formation of biofilms, production of valuable biomolecules, and growth on renewable substrates. Identification of the genetic networks responsible for these phenotypes represents the enabling technology to enhance or tune strain performance.

Siderophore synthesis pathways have been identified in many organisms through both experiments and bioinformatics (Hider and Kong 2010; Garber et al. 2020, 2021). Several pathways have been expressed and further engineered in the model organism *Escherichia coli*, with the goal of producing siderophores as therapeutic antibiotics (Fujita and Sakai 2013; Fujita et al. 2018). To our knowledge, there have been limited efforts to enhance the production of endogenous siderophores, to date. Siderophores are typically produced in bacteria by nonribosomal peptide synthesis and secretion, which are both resource- and energy-intensive processes (Hider and Kong 2010). Enhanced production may require the constitutive expression of siderophore biosynthesis systems both endogenously and heterologously, metabolic engineering for increased

availability of siderophore precursors, or alterations of the bacterial secretion complex. Although transcriptomic studies have been performed on siderophore-producing organisms under iron-limiting conditions (Manck et al. 2020), the biological bottleneck for the production of siderophores is sparingly known and often complex. For example, *Staphylococcus aureus* can produce both staphyloferrin A (SA) or B (SB), both of which are citrate-based siderophores; down-regulation of the tricarboxylic acid (TCA) during Fe limitation limits the production of SA because of decreased citrate synthase activity. An alternate citrate synthase up-regulated by Fe limitation supports the synthesis of SB, however. This system provides a natural inspiration to transcriptionally and metabolically maintain enhanced siderophore synthesis (Sheldon et al. 2014). Enhanced or constitutive secretion of siderophores could shed light on the possible upper limit of siderophore-mediated silicate dissolution and contribute to the broader understanding of engineered synthesis and production of complex natural products. There is also the point that siderophores are not a chemical monolith, but rather a functional category composed of highly diverse chemical structures and backbones, which influences their bioavailability, chemical behavior in water, and photoreactivity (Barbeau et al. 2003). Exploring the specific interactions between siderophores and olivine will better inform possible designer siderophores that are both metabolically cheap to produce and optimized for catalyzing dissolution.

Genetic engineering may also enhance other mechanisms of mineral dissolution, such as biofilm formation, citrate production, and local acidification. Biofilm formation in the context of pathogenesis, for example, has been studied extensively both genetically and physiologically, resulting in a range of therapeutics and drugs that disrupt biofilm formation (Jiang et al. 2020). Biofilms could be used industrially for olivine dissolution and other applications such as microbe–electrode interactions, but engineering and controls of film formation and architecture remain difficult (Angelaalincy et al. 2018; Mukherjee et al. 2020).

Many of the phenotypes described here, such as siderophore production and biofilm formation, are difficult to engineer rationally because of their complex genetics. In the system proposed here, however, microbial growth is coupled to the harvesting of iron from olivine and other silicate minerals. It may be possible, therefore, to perform large-scale selection of mutants for enhanced harvesting of nutrients from minerals. High-throughput selections or screens could reveal key targets for rational genetic engineering and identify high-performing gain-of-function mutations that have immediate industrial utility.

Bottom-up genetic engineering and synthetic biology can improve the performance of coculture or consortia systems (Lawson et al. 2019). For operation at large scales, the stability of a multicomponent living system is challenging, especially in complex, nonsterile mediums. Mutual dependencies between microbial organisms have been extensively described in natural systems and applied to synthetic systems at laboratory scale (Schink 2002; Goers et al. 2014; Chen et al. 2019). To date, synthetic microbial consortia have not been implemented at industrial scale, however. The olivine dissolution system described here, specifically, offers a novel solution for synthetic mutual dependency.

Finally, metabolic engineering and synthetic biology have been extensively applied to synthesize valuable biomolecules at high titers in engineered hosts. These include biomass components such as lipids, complex natural products, biofuels, and medicines (Chemler and Koffas 2008; Courchesne et al. 2009; Adrio and Demain 2010; Choi et al. 2020). Similar techniques are applied to enhance the growth of microorganisms on renewable feedstocks, such as lignocellulose, or plastic polymers as discussed earlier (Huang et al. 2014). One additional microbial application of olivine dissolution is the uptake and concentration of valuable trace metals. Metal uptake by bacteria has been studied in the context of heavy metal remediation (Gadd 1990; Valls and Lorenzo 2002) and plant–microbe interactions (Kidd et al. 2017). A system for microbial dissolution of minerals provides a uniquely useful platform with which to study the uptake and harvesting of valuable metals.

Utilization of genetically engineered organisms for large-scale mineral dissolution requires consideration of the ethical and ecological impacts. One advantage of the proposed system is the potential for physical containment on top of genetic controls. Furthermore, engineering controls may protect against unwanted mutation or ecosystem proliferation. Recently reported organisms with a reduced genetic code may be less likely to acquire natural genetic material that could confer new phenotypes and contribute synthetic genetic material to the natural gene pool (Lau et al. 2017; Robertson et al. 2021). Novel kill switches and synthetic auxotrophy may also restrict the activity and replication of engineered organisms to controlled environments (Stirling et al. 2018, 2020; Whitford et al. 2018; Stirling and Silver 2020). In 2022, the Engineering Biology Research Consortium released a road map for the application of synthetic biology to climate and sustainability that urges the identification, assessment, and open communication of benefits and risks associated with research efforts (Chen et al. 2022). Early examples of deployment of engineered organisms such as engineered crops with resistance to pests or drought provide a framework within which new technologies may be evaluated (Scheben and Edwards 2017; Smyth 2020).

In short, physical, chemical, and genetic conditions may be engineered to optimize metabolic functions, as well as the thermodynamics and kinetics of mineral dissolution and carbon capture. Such cultivation systems can be operated within a broader existing economic and industrial infrastructure, in which they can act not only as systems of large-scale carbon capture, but also offer sustainable means of waste management and/or production of valuable by-products.

Considerations for Mineral Feedstocks and Industrial Cultivation

Large-scale (Mt/Gt) CDR reactors will require large inputs of water, nutrients, and mineral feedstock and will require downstream processing to recover valuable added products. Industrial infrastructure and scientific knowledge already exist to implement such reactors at scale. Combined, the mining and construction sectors produce >8 Gt/yr of mineral waste materials with CDR potential >8 Gt/yr (Renforth 2019; Bullock et al. 2022). The remaining questions then relate to finding the upper limits of dissolution and carbon capture rates that such reactors can offer, while identifying the operating conditions that will sustain these at scale.

Some of the key questions that remain include identifying the conditions for maximal mineral dissolution and carbon capture rates. Reaching these maxima will involve experimenting with a wide range of physical and chemical constraints, as these help manage the baseline thermodynamic environment of mineral dissolution. As discussed, silicate dissolution rates can be difficult to constrain over periods of years. There is a significant bias in the literature, within which quantified measures of dissolution have focused on short time periods of hours to days (Fig. 2; Oelkers et al. 2018), which may not be representative of long-term steady state dissolution rates. Figure 2 shows a modeled dissolution rate derived from the first 300 h of dissolution versus empirical dissolution data over several weeks. Moreover, characterizing mineral dissolution over months can, however, be challenging, as dissolution can be nonstoichiometric, and different proxies (i.e., dissolved silicon vs. nickel) can give different quantified estimates of dissolution rates (Montserrat et al. 2017).

Additionally, a comprehensive characterization of mineral substrates and their geochemistry is key. As discussed, a mineral's dissolution rate and potential to generate alkalinity are critical characteristics in determining the carbon capture potential. In this respect, ultramafic minerals such as olivine and serpentine are of great interest (Bullock et al. 2022). However, associated reactions can negatively impact carbon capture efficiency, such as the oxidation of iron. As a major component of olivine and many ultramafic minerals, ferrous iron will oxidize readily during mineral dissolution, a reaction that releases free protons and neutralizes the alkalinity generation congruent to silicate dissolution. Choosing ultramafic minerals with low iron content would be important in abiotic conditions. However, in the context of bioreactors that promote

microbial iron acquisition, it is unclear what iron concentration will maximize dissolution rates. Furthermore, the processes involved in the microbial acquisition of iron (i.e., ligand exchange reactions between siderophores and ferric iron, ferric reduction involved in iron cellular absorption, along with photoreduction of ligand-bound iron) are all proton-sensitive reactions and can impact pH and thus alkalinity generation to some extent (Barbeau et al. 2002; Dhungana and Crumbliss 2005). Characterizing this process will help constrain system design for maximal carbon capture.

Another critical component in estimating net CO_2 fluxes relates to the fate of carbon held in the organic pool. Characterizing the fluxes and fate of carbon across processes like biological assimilation, biofilms, mineral dissolution, and secondary precipitation in a dense environment is a challenging yet exciting and meaningful scientific endeavor. Although the single most important metric for a bioreactor might be net carbon fluxes and associated carbon capture rates, it will be important to characterize the fate of carbon and all its forms. This implies steep geochemical and metabolic gradients, forcing a wide range of reactions that can impact the form and fate of carbon. The main forms of carbon flowing out of bioreactors will be dissolved as well as particulate organic and inorganic phases. Being able to account for all their relative abundances will be valuable, not only as an additional means to measure carbon capture and potentially useful biomolecule production, but also to inform on strategies around reactor design, engineering, and permanent carbon storage.

Inorganic carbon fluxes are highly dependent on a system's carbonate saturation and precipitation as secondary mineral phases (Fuhr et al. 2022). As per Equation 1, the most efficient form of carbon capture is in the form of bicarbonate (also referred to as dissolution trapping). An ideal bioreactor system should therefore maintain conditions below carbonate and silicate saturation to limit the precipitation of secondary minerals. Although the precipitation of carbonates still produces a net sink of stored carbon (also referred to as mineral trapping or carbonation), the acidity simultaneously generated in this reac-

Cite this article as *Cold Spring Harb Perspect Biol* doi: 10.1101/cshperspect.a041673

tion decreases the efficacy of the process significantly (Fuhr et al. 2022). The flexibility of an engineered system offers the possibility to promote and effectively manage carbonate precipitation as a form of long-term storage.

Hence, a comprehensive characterization of accelerated mineral dissolution geochemistry under a wide range of conditions will be critical to maximize carbon capture, and the forms of carbon must be fully characterized to best manage its permanent storage. Carbonate precipitates can be easy to store permanently because they are relatively unreactive in the natural environment. Bicarbonate and DIC, however, need to be stored in vast, well-buffered basins such as the world's oceans to remain stable and not outgas CO_2 back into the atmosphere. Organic forms of carbon (particulate and dissolved) will need to be stored in an environment where it will avoid re-oxidation, which might involve specific containment strategies.

Techno-Economic Viability

The total CDR potential of known, global reserves of ultramafic minerals lies in the range of several tens of thousands of $GtCO_2$, orders of magnitude greater than what is required to sequester anthropogenic CO_2 emissions (Lackner 2002; Power et al. 2013). Although geochemical CDR thus offers tremendous potential, implementing it at scale carries economic constraints that must be taken into account. Techno-economic analyses of large-scale reactors are critical in guiding research and development, highlighting constraints that are not always obvious in a laboratory or small-scale setting.

By estimating the dimensions and requirements of a reactor at scale, one can infer the costs and logistical constraints of such an operation. Namely, these will include the costs of building the reactors, excavating and grinding the ultramafic minerals, the costs of fertilizers and other substrates, accessing and treating the water used in reactors, and total energy requirements. The economic potential of carbon capture also offers a unique opportunity for industries that use microbial catalysis as a process of chemical engineering. Some industries have been extremely

successful (i.e., fermentation and the alcoholic beverages industry), and some have struggled to remain competitive (i.e., algae farms and biofuels industry) (Hoffman et al. 2017). Carbon capture reactors, however, sell, at their core, carbon credits. The recovery of additional, valuable by-products like metals could serve to subsidize the cost of carbon capture and will further help integrate CDR efforts into the broader economy, by providing a source of valuable commodities to other sectors including agriculture and aquaculture, pharmaceuticals, and other industries. This represents an unprecedented opportunity to connect the accelerated weathering of the ultramafic minerals to rapidly growing carbon markets projected to surpass $1 trillion in 2050, as net-zero pledges from countries and companies continue to skyrocket (Liu et al. 2022).

Thus far, our techno-economic analyses show that industrial-scale bioreactor systems for enhanced weathering are comparable to common water treatment facilities and can thus benefit from existing infrastructure and engineering capabilities. With the scientific proof of concept already established (Berghe et al. 2021), our analyses indicate that methodological and technological improvements to scale bioreactors as described above have the potential to achieve net cost of carbon capture (including capex, opex, energy and carbon costs, substrate inputs) of ~$100/tonne. Because of these valuable, additional products, this process not only has the potential to be inherently profitable, but could be considered a simultaneous carbon capture and mining and resource extraction process.

CONCLUSION

Because of inaction for broad decarbonization coupled to continuous anthropogenic GHG emissions, there is a global consensus that large-scale CDR will be needed to remove $GtCO_2$ over the next several decades. CDR will also be necessary to remove further CO_2 emissions required for the transition to a decarbonized, green economy. Accelerating Earth's inorganic carbon cycle through the weathering of silicate minerals like olivine offers a promising path to removing and storing gigatonnes of atmospheric CO_2. Because of the sheer amount of CO_2 that needs to be sequestered, however, multiple weathering approaches must be pursued simultaneously. Although grinding minerals into sand-sized particles can accelerate weathering rates of olivine minerals in coastal environments (i.e., coastal enhanced weathering), half-lives of olivine grains are on the order of decades. Hence, we suggest that microbial catalysis to accelerate mineral dissolution rates must be explored alongside other CDR methods. We have shown that microbes have the potential to significantly increase weathering rates leading to olivine grain half-lives of several years. By using micronutrients like Fe in olivine, microbes can use these mineral substrates for exponential growth, thereby offering a path to scale the cultivation of microbiomes growing on silicate minerals. Indeed, other industries have successfully cultivated microbes to perform specific, valuable functions including wastewater treatment and fermentation. Further research to examine the biological mechanisms of microbial–mineral interactions, biofilm formation, and large-scale cultivation will provide critical insights that can galvanize global efforts in silicate dissolution for CDR. It will also create a step change in our understanding of fundamental microbial processes particularly in biofilm formation and the synthesis of complex natural products, while also opening the door to novel approaches to engineer microbiomes. These research endeavors naturally integrate the fields of CDR, synthetic biology, geochemistry, climate science, engineering, and economics and will provide new models that can yield various outputs from the sustainable biosynthesis of valuable products to the removal and long-term storage of excess atmospheric CO_2. Although many of these fundamental processes have been happening worldwide for billions of years, we are only taking the initial steps along the pathway toward understanding how these mechanisms can help turn the tide on the climate crisis.

ACKNOWLEDGMENTS

This article has been made freely available online by generous financial support from the Bill and Melinda Gates Foundation. We particularly want to acknowledge the support and encouragement of Rodger Voorhies, president, Global Growth and Opportunity Fund.

REFERENCES

Adrio JL, Demain AL. 2010. Recombinant organisms for production of industrial products. *Bioeng Bugs* 1: 116–131. doi:10.4161/bbug.1.2.10484

Angelaalincy MJ, Krishnaraj RN, Shakambari G, Ashokkumar B, Kathiresan S, Varalakshmi P. 2018. Biofilm engineering approaches for improving the performance of microbial fuel cells and bioelectrochemical systems. *Front Energy Res* 6: 63. doi:10.3389/fenrg.2018.00063

Arora NK, Mishra I. 2021. COP26: more challenges than achievements. *Environ Sustain* 4: 585–588. doi:10.1007/s42398-021-00212-7

Barbeau K, Zhang G, Live DH, Butler A. 2002. Petrobactin, a photoreactive siderophore produced by the oil-degrading marine bacterium *Marinobacter hydrocarbonoclasticus*. *J Am Chem Soc* 124: 378–379. doi:10.1021/ja0119088

Barbeau K, Rue EL, Trick CG, Bruland KW, Butler A. 2003. Photochemical reactivity of siderophores produced by marine heterotrophic bacteria and cyanobacteria based on characteristic Fe(III) binding groups. *Limnol Oceanogr* 48: 1069–1078. doi:10.4319/lo.2003.48.3.1069

Beerling DJ, Kantzas EP, Lomas MR, Wade P, Eufrasio RM, Renforth P, Sarkar B, Andrews MG, James RH, Pearce CR, et al. 2021. Potential for large-scale CO_2 removal via enhanced rock weathering with croplands. *Nature* 583: 242–248. doi:10.1038/s41586-020-2448-9

Berghe MVD, Merino N, Nealson KH, West AJ. 2021. Silicate minerals as a direct source of limiting nutrients: siderophore synthesis and uptake promote ferric iron bioavailability from olivine and microbial growth. *Geobiology* 19: 618–630. doi:10.1111/gbi.12457

Brophy JAN, Triassi AJ, Adams BL, Renberg RL, Stratis-Cullum DN, Grossman AD, Voigt CA. 2018. Engineered integrative and conjugative elements for efficient and inducible DNA transfer to undomesticated bacteria. *Nat Microbiol* 3: 1043–1053. doi:10.1038/s41564-018-0216-5

Bullock LA, Yang A, Darton RC. 2022. Kinetics-informed global assessment of mine tailings for CO_2 removal. *Sci Total Environ* 808: 152111. doi:10.1016/j.scitotenv.2021.152111

Campbell JS, Foteinis S, Furey V, Hawrot O, Pike D, Aeschlimann S, Maesano CN, Reginato PL, Goodwin DR, Looger LL, et al. 2022. Geochemical negative emissions technologies. Part I: Review. *Front Clim* 4: 879133. doi:10.3389/fclim.2022.879133

Cawley GC. 2011. On the atmospheric residence time of anthropogenically sourced carbon dioxide. *Energy Fuels* 25: 5503–5513. doi:10.1021/ef200914u

Chemler JA, Koffas MA. 2008. Metabolic engineering for plant natural product biosynthesis in microbes. *Curr Opin Biotechnol* 19: 597–605. doi:10.1016/j.copbio.2008.10.011

Chen T, Zhou Y, Lu Y, Zhang H. 2019. Advances in heterologous biosynthesis of plant and fungal natural products by modular co-culture engineering. *Biotechnol Lett* **41:** 27–34. doi:10.1007/s10529-018-2619-z

Chen S, Aurand E, Köpke M. 2022. Engineering biology for climate & sustainability: a research roadmap for a cleaner future. https://roadmap.ebrc.org/wp-content/uploads/2022/09/Engineering-Biology-for-Climate-and-Sustainability-A-Research-Roadmap-for-a-Cleaner-Future_20September2022.pdf

Choi KR, Jiao S, Lee SY. 2020. Metabolic engineering strategies toward production of biofuels. *Curr Opin Chem Biol* **59:** 1–14. doi:10.1016/j.cbpa.2020.02.009

Courchesne NMD, Parisien A, Wang B, Lan CQ. 2009. Enhancement of lipid production using biochemical, genetic and transcription factor engineering approaches. *J Biotechnol* **141:** 31–41. doi:10.1016/j.jbiotec.2009.02.018

Dhungana S, Crumbliss AL. 2005. Coordination chemistry and redox processes in siderophore-mediated iron transport. *Geomicrobiol J* **22:** 87–98. doi:10.1080/01490450590945870

Diallo MS, Kotte MR, Cho M. 2015. Mining critical metals and elements from seawater: opportunities and challenges. *Environ Sci Technol* **49:** 9390–9399. doi:10.1021/acs.est.5b00463

Dreyfus GB, Xu Y, Shindell DT, Zaelke D, Ramanathan V. 2022. Mitigating climate disruption in time: a self-consistent approach for avoiding both near-term and long-term global warming. *Proc Natl Acad Sci* **119:** e2123536119. doi:10.1073/pnas.2123536119

Fuhr M, Geilert S, Schmidt M, Liebetrau V, Vogt C, Ledwig B, Wallmann K. 2022. Kinetics of olivine weathering in seawater: an experimental study. *Front Clim* **4:** 831587. doi:10.3389/fclim.2022.831587

Fujita MJ, Sakai R. 2013. Heterologous production of desferrioxamines with a fusion biosynthetic gene cluster. *Biosci Biotechnol Biochem* **77:** 2467–2472. doi:10.1271/bbb.130597

Fujita MJ, Goto Y, Sakai R. 2018. Cloning of the bisucaberin B biosynthetic gene cluster from the marine bacterium *Tenacibaculum mesophilum*, and heterologous production of bisucaberin B. *Mar Drugs* **16:** 342. doi:10.3390/md16090342

Gadd GM. 1990. Heavy metal accumulation by bacteria and other microorganisms. *Experientia* **46:** 834–840. doi:10.1007/BF01935534

Garber AI, Nealson KH, Okamoto A, McAllister SM, Chan CS, Barco RA, Merino N. 2020. Fegenie: a comprehensive tool for the identification of iron genes and iron gene neighborhoods in genome and metagenome assemblies. *Front Microbiol* **11:** 37. doi:10.3389/fmicb.2020.00037

Garber AI, Cohen AB, Nealson KH, Ramírez GA, Barco RA, Enzingmüller-Bleyl TC, Gehringer MM, Merino N. 2021. Metagenomic insights into the microbial iron cycle of subseafloor habitats. *Front Microbiol* **12:** 667944. doi:10.3389/fmicb.2021.667944

Goers L, Freemont P, Polizzi KM. 2014. Co-culture systems and technologies: taking synthetic biology to the next level. *J R Soc Interface* **11:** 20140065. doi:10.1098/rsif.2014.0065

Guerinot ML, Yi Y. 1994. Iron: nutritious, noxious, and not readily available. *Plant Physiol* **104:** 815–820. doi:10.1104/pp.104.3.815

Hall-Stoodley L, Costerton JW, Stoodley P. 2004. Bacterial biofilms: from the natural environment to infectious diseases. *Nat Rev Microbiol* **2:** 95–108. doi:10.1038/nrmicro821

Hartmann J, West AJ, Renforth P, Köhler P, Rocha CLDL, Wolf-Gladrow DA, Dürr HH, Scheffran J. 2013. Enhanced chemical weathering as a geoengineering strategy to reduce atmospheric carbon dioxide, supply nutrients, and mitigate ocean acidification. *Rev Geophys* **51:** 113–149. doi:10.1002/rog.20004

He J, Tyka MD. 2022. Limits and CO$_2$ equilibration of near-coast alkalinity enhancement. *Egusphere* **2022:** 1–26.

Hider RC, Kong X. 2010. Chemistry and biology of siderophores. *Nat Prod Rep* **27:** 637–657. doi:10.1039/b906679a

Hoffman J, Pate RC, Drennen T, Quinn JC. 2017. Techno-economic assessment of open microalgae production systems. *Algal Res* **23:** 51–57. doi:10.1016/j.algal.2017.01.005

Huang GL, Anderson TD, Clubb RT. 2014. Engineering microbial surfaces to degrade lignocellulosic biomass. *Bioengineered* **5:** 96–106. doi:10.4161/bioe.27461

Hutchins DA, Boyd PW. 2016. Marine phytoplankton and the changing ocean iron cycle. *Nat Clim Change* **6:** 1072–1079. doi:10.1038/nclimate3147

Jiang Y, Geng M, Bai L. 2020. Targeting biofilms therapy: current research strategies and development hurdles. *Microorganisms* **8:** 1222. doi:10.3390/microorganisms8081222

Kidd PS, Álvarez-López V, Becerra-Castro C, Cabello-Conejo M, Prieto-Fernández Á. 2017. Chapter three potential role of plant-associated bacteria in plant metal uptake and implications in phytotechnologies. *Adv Bot Res* **83:** 87–126. doi:10.1016/bs.abr.2016.12.004

Lackner KS. 2002. Carbonate chemistry for sequestering fossil carbon. *Annu Rev Energy Environ* **27:** 193–232. doi:10.1146/annurev.energy.27.122001.083433

Lau YH, Stirling F, Kuo J, Karrenbelt MAP, Chan YA, Riesselman A, Horton CA, Schäfer E, Lips D, Weinstock MT, et al. 2017. Large-scale recoding of a bacterial genome by iterative recombineering of synthetic DNA. *Nucleic Acids Res* **45:** 6971–6980. doi:10.1093/nar/gkx415

Lawson CE, Harcombe WR, Hatzenpichler R, Lindemann SR, Löffler FE, O'Malley MA, Martín HG, Pfleger BF, Raskin L, Venturelli OS, et al. 2019. Common principles and best practices for engineering microbiomes. *Nat Rev Microbiol* **17:** 725–741. doi:10.1038/s41579-019-0255-9

Ledyard KM, Butler A. 1997. Structure of putrebactin, a new dihydroxamate siderophore produced by *Shewanella putrefaciens*. *J Biol Inorg Chem* **2:** 93–97. doi:10.1007/s007750050110

Lee ED, Aurand ER, Friedman DC; Engineering Biology Research Consortium Microbiomes Roadmapping Working Group. 2020. Engineering microbiomes—looking ahead. *ACS Synth Biol* **9:** 3181–3183. doi:10.1021/acssynbio.0c00558

Lenton TM, Rockström J, Gaffney O, Rahmstorf S, Richardson K, Steffen W, Schellnhuber HJ. 2019. Climate tipping points—too risky to bet against. *Nature* **575**: 592–595. doi:10.1038/d41586-019-03595-0

Liu Q, Chen Z, Xiao SX. 2022. A theory of carbon currency. *Fundam Res* **2**: 375–383. doi:10.1016/j.fmre.2022.02.007

Lunstrum A, Berghe MVD, Bian X, John S, Nealson K, West AJ. 2023. Bacterial use of siderophores increases olivine dissolution rates by nearly an order of magnitude. *Geochem Perspect Lett* **25**: 51–55. doi:10.7185/geochemlet.2315

Manck LE, Espinoza JL, Dupont CL, Barbeau KA. 2020. Transcriptomic study of substrate-specific transport mechanisms for iron and carbon in the marine copiotroph *Alteromonas macleodii*. *Msystems* **5**: e00070-20. doi:10.1128/mSystems.00070-20

Manck LE, Park J, Tully BJ, Poire AM, Bundy RM, Dupont CL, Barbeau KA. 2022. Petrobactin, a siderophore produced by *Alteromonas*, mediates community iron acquisition in the global ocean. *ISME J* **16**: 358–369. doi:10.1038/s41396-021-01065-y

Meysman FJR, Montserrat F. 2017. Negative CO_2 emissions via enhanced silicate weathering in coastal environments. *Biol Lett* **13**: 20160905. doi:10.1098/rsbl.2016.0905

Middelburg JJ, Soetaert K, Hagens M. 2020. Ocean alkalinity, buffering and biogeochemical processes. *Rev Geophys* **58**: e2019RG000681. doi:10.1029/2019RG000681

Montserrat F, Renforth P, Hartmann J, Leermakers M, Knops P, Meysman FJR. 2017. Olivine dissolution in seawater: implications for CO_2 sequestration through enhanced weathering in coastal environments. *Environ Sci Technol* **51**: 3960–3972. doi:10.1021/acs.est.6b05942

Mukherjee M, Zaiden N, Teng A, Hu Y, Cao B. 2020. Shewanella biofilm development and engineering for environmental and bioenergy applications. *Curr Opin Chem Biol* **59**: 84–92. doi:10.1016/j.cbpa.2020.05.004

National Academies of Sciences Engineering and Medicine. 2022. *A research strategy for ocean-based carbon dioxide removal and sequestration*. The National Academies Press, Washington, DC.

Oelkers EH, Benning LG, Lutz S, Mavromatis V, Pearce CR, Plümper O. 2015. The efficient long-term inhibition of forsterite dissolution by common soil bacteria and fungi at Earth surface conditions. *Geochim Cosmochim Acta* **168**: 222–235. doi:10.1016/j.gca.2015.06.004

Oelkers EH, Declercq J, Saldi GD, Gislason SR, Schott J. 2018. Olivine dissolution rates: a critical review. *Chem Geol* **500**: 1–19. doi:10.1016/j.chemgeo.2018.10.008

Pokrovsky OS, Shirokova LS, Benezeth P, Schott J, Golubev SV. 2009. Effect of organic ligands and heterotrophic bacteria on wollastonite dissolution kinetics. *Am J Sci* **309**: 731–772. doi:10.2475/08.2009.05

Power IM, Wilson SA, Dipple GM. 2013. Serpentinite carbonation for CO_2 sequestration. *Elements* **9**: 115–121. doi:10.2113/gselements.9.2.115

Renforth P. 2019. The negative emission potential of alkaline materials. *Nat Commun* **10**: 1401. doi:10.1038/s41467-019-09475-5

Renforth P, Henderson G. 2017. Assessing ocean alkalinity for carbon sequestration. *Rev Geophys* **55**: 636–674. doi:10.1002/2016RG000533

Rimstidt JD, Brantley SL, Olsen AA. 2012. Systematic review of forsterite dissolution rate data. *Geochim Cosmochim Acta* **99**: 159–178. doi:10.1016/j.gca.2012.09.019

Robertson WE, Funke LFH, de la Torre D, Fredens J, Elliott TS, Spinck M, Christova Y, Cervettini D, Böge FL, Liu KC, et al. 2021. Sense codon reassignment enables viral resistance and encoded polymer synthesis. *Science* **372**: 1057–1062. doi:10.1126/science.abg3029

Rogers P, Chen JS, Zedwick MJ. 2013. Organic acid and solvent production: acetic, lactic, gluconic, succinic, and polyhydroxyalkanoic acids. In *The prokaryotes, applied bacteriology and biotechnology* (ed. Rosenberg E, DeLong EF, Lory S, et al.). Springer, Berlin.

Rosso JJ, Rimstidt JD. 2000. A high resolution study of forsterite dissolution rates. *Geochim Cosmochim Acta* **64**: 797–811. doi:10.1016/S0016-7037(99)00354-3

Scheben A, Edwards D. 2017. Genome editors take on crops. *Science* **355**: 1122–1123. doi:10.1126/science.aal4680

Schink B. 2002. Synergistic interactions in the microbial world. *Antonie Van Leeuwenhoek* **81**: 257–261. doi:10.1023/A:1020579004534

Schott J, Berner RA. 1983. X-ray photoelectron studies of the mechanism of iron silicate dissolution during weathering. *Geochim Cosmochim Acta* **47**: 2233–2240. doi:10.1016/0016-7037(83)90046-7

Sheldon JR, Marolda CL, Heinrichs DE. 2014. TCA cycle activity in *Staphylococcus aureus* is essential for iron-regulated synthesis of staphyloferrin A, but not staphyloferrin B: the benefit of a second citrate synthase. *Mol Microbiol* **92**: 824–839. doi:10.1111/mmi.12593

Smyth SJ. 2020. Regulatory barriers to improving global food security. *Glob Food Secur* **26**: 100440. doi:10.1016/j.gfs.2020.100440

Stirling F, Silver PA. 2020. Controlling the implementation of transgenic microbes: are we ready for what synthetic biology has to offer? *Mol Cell* **78**: 614–623. doi:10.1016/j.molcel.2020.03.034

Stirling F, Bitzan L, O'Keefe S, Redfield E, Oliver JWK, Way J, Silver PA. 2018. Rational design of evolutionarily stable microbial kill switches. *Mol Cell* **72**: 395. doi:10.1016/j.molcel.2018.10.002

Stirling F, Naydich A, Bramante J, Barocio R, Certo M, Wellington H, Redfield E, O'Keefe S, Gao S, Cusolito A, et al. 2020. Synthetic cassettes for pH-mediated sensing, counting, and containment. *Cell Rep* **30**: 3139–3148.e4. doi:10.1016/j.celrep.2020.02.033

Suzuki M, Tachibana Y, Kasuya K. 2021. Biodegradability of poly(3-hydroxyalkanoate) and poly(ε-caprolactone) via biological carbon cycles in marine environments. *Polym J* **53**: 47–66. doi:10.1038/s41428-020-00396-5

Torres MA, West AJ, Nealson K. 2014. Microbial acceleration of olivine dissolution via siderophore production. *Procedia Earth Planet Sci* **10**: 118–122. doi:10.1016/j.proeps.2014.08.041

Torres MA, Dong S, Nealson KH, West AJ. 2019. The kinetics of siderophore-mediated olivine dissolution. *Geobiology* **17**: 401–416. doi:10.1111/gbi.12332

Valls M, Lorenzo V. 2002. Exploiting the genetic and biochemical capacities of bacteria for the remediation of heavy metal pollution. *Fems Microbiol Rev* **26**: 327–338. doi:10.1016/S0168-6445(02)00114-6

Van Den Berghe M, Merino N, Nealson KH, West AJ. 2021. Silicate minerals as a direct source of limiting nutrients: siderophore synthesis and uptake promote ferric iron bioavailability from olivine and microbial growth. *Geobiology* **19:** 618–630. doi:10.1111/gbi.12457

Whitford CM, Dymek S, Kerkhoff D, März C, Schmidt O, Edich M, Droste J, Pucker B, Rückert C, Kalinowski J. 2018. Auxotrophy to Xeno-DNA: an exploration of combinatorial mechanisms for a high-fidelity biosafety system for synthetic biology applications. *J Biol Eng* **12:** 13. doi:10.1186/s13036-018-0105-8

Williamson P, Wallace DWR, Law CS, Boyd PW, Collos Y, Croot P, Denman K, Riebesell U, Takeda S, Vivian C. 2012. Ocean fertilization for geoengineering: a review of effectiveness, environmental impacts and emerging governance. *Process Saf Environ Prot* **90:** 475–488. doi:10.1016/j.psep.2012.10.007

Włodarczyk A, Selão TT, Norling B, Nixon PJ. 2020. Newly discovered *Synechococcus* sp. PCC 11901 is a robust cyanobacterial strain for high biomass production. *Commun Biol* **3:** 215. doi:10.1038/s42003-020-0910-8

Photosynthesis 2.0: Realizing New-to-Nature CO$_2$-Fixation to Overcome the Limits of Natural Metabolism

Tobias J. Erb

Max Planck Society, Germany, Department for Biochemistry & Synthetic Metabolism, Center for Synthetic Microbiology (SYNMIKRO) & Max Planck Institute for Terrestrial Microbiology, 35043 Marburg, Germany

Correspondence: toerb@mpi-marburg.mpg.de

Synthetic biology provides opportunities to realize new-to-nature CO$_2$-fixation metabolisms to overcome the limitations of natural photosynthesis. Two different strategies are currently being pursued: One is to realize engineered plants that feature carbon-neutral or carbon-negative (i.e., CO$_2$-fixing) photorespiration metabolism, such as the tatronyl-CoA (TaCo) pathway, to boost CO$_2$-uptake rates of photosynthesis between 20% and 60%. Another (arguably more radical) is to create engineered plants in which natural photosynthesis is fully replaced by an alternative CO$_2$-fixation metabolism, such as the CETCH cycle, which carries the potential to improve CO$_2$ uptake rates between 20% and 200%. These efforts could revolutionize plant engineering by expanding the capabilities of plant metabolism beyond the constraints of natural evolution to create highly improved crops addressing the challenges of climate change in the future.

Global warming, caused by anthropogenic emissions of greenhouse gases (in particular CO$_2$), has increased global temperatures by 1.1°C compared to preindustrial records, and the global average temperature is expected to rise a further 2.5°C to 4°C by the end of the century. This increase in global warming causes more extreme weather events, such as droughts, heat waves, fires, and flooding.

Agriculture will need to cope more effectively with these events, while it also needs to feed an ever-growing world population that is projected to reach 9.7 billion in 2050. Taking all this together, future crops will need to be of higher resilience and productivity, plus serve as more efficient carbon sinks to help mitigate climate change on a global scale (World Population Prospects 2019). However, photosynthetic CO$_2$-fixation has its natural limits, which cannot be overcome simply through state-of-the art plant engineering or breeding methods (Erb and Zarzycki 2016; Wurtzel et al. 2019).

In contrast, the emerging field of synthetic biology provides completely new and exciting opportunities to fundamentally re-think, re-design, and re-construct photosynthesis to improve the CO$_2$-uptake capabilities of plants. Some of these new-to-nature solutions have the potential to increase CO_2 uptake in crops between 20% and as much as 200%, which would translate to an increased uptake of 2 to 19 Gt C per year on a global scale.

RUBISCO IS A BOTTLENECK IN PHOTOSYNTHETIC CO_2 FIXATION

Climate change has become one of the key challenges of the twenty-first century. While different technical carbon-capturing solutions are currently being explored, Nature itself already provides the blueprint for sustainable capture and conversion at a global scale: photosynthesis. Photosynthesis captures and converts 100 Gt C (equivalent to 400 Gt CO_2) per year, which is almost ten times more than the annual anthropogenic CO_2 emissions of about 10 Gt C (40 Gt CO_2) per year. Yet, natural photosynthesis is unable to compensate for these human-made CO_2 emissions and will increasingly suffer from a rapidly changing environment, which will negatively impact agricultural productivity.

Thus, new approaches are required to overcome the natural limits of photosynthesis. Increasing photosynthetic efficiency will directly improve plant productivity and CO_2-uptake, thereby not only leading to increased plant biomass productivity, but also helping to mitigate the effects of climate change due to an improved carbon uptake.

Unfortunately, classical and new breeding technologies are not able to address this fundamental challenge, because they are still bound to the naturally existing solution space and its natural limits. This relates especially to the "dark reaction" of photosynthesis, the process in which CO_2 is captured and converted, also known as the Calvin–Benson–Bassham (CBB) cycle, and its key enzyme, the CO_2-fixing enzyme ribulose-1,5-bisphosphate carboxylase/oxygenase (rubisco) (Bassham et al. 1954).

Even though rubisco is at the very core of the CBB cycle, the enzyme has a very low turnover rate. In an average plant, rubisco turns over only 5–10 CO_2 molecules per second, which is more than an order of magnitude slower than most enzymes in central carbon metabolism. This makes carbon capture through rubisco one of the (if not *the*) limiting steps in photosynthesis. In addition, rubisco also shows a side reaction with oxygen (O_2), which leads to an oxygenation (instead of a carboxylation) of ribulose-1,5-bisphosphate. This O_2-fixing activity causes the process of photorespiration, which releases up to almost 30% of the fixed carbon in photosynthesis, directly lowering photosynthetic yield (Erb and Zarzycki 2016; Walker et al. 2016).

Several studies have focused on improving either catalytic activity of rubisco or the enzyme's CO_2 specificity (e.g., by suppressing the oxygenase reaction). Yet, all these efforts have shown that catalytic activity and specificity in rubisco are most likely inversely linked with each other, suggesting that the enzyme has evolved along a Pareto front, in which one parameter (activity) cannot be changed without negatively affecting the other (specificity). Recent evolutionary studies also seem to support this view (Schulz et al. 2022). In summary, even though rubisco is responsible for more than 90% of CO_2-fixation in the global carbon cycle, it is a limiting factor in photosynthesis that cannot be easily engineered for improved carbon capture.

SYNTHETIC BIOLOGY TO REALIZE HIGHLY EFFICIENT CO_2-FIXATION PATHWAYS THAT NATURE HAS NOT INVENTED

Notably, the CBB cycle is not the only CO_2-fixation pathway and rubisco is not the only CO_2-fixing enzyme that nature has invented during evolution. Over the last few years, eight additional, so-called, "alternative" CO_2-fixation pathways have been discovered in different microbes. Not all of these pathways rely on rubisco, some exploit other CO_2-fixing enzymes (Berg 2011; Könneke et al. 2014; Bierbaumer et al. 2023). Although these pathways (still) lag behind the productivity of photosynthetic CO_2 fixation, together they fix several Gt CO_2 per year, which shows that other solutions beyond the CBB cycle and rubisco are in principle possible.

However, while these eight "alternative" CO_2-fixation pathways are a great example for the creativity of evolution, they actually represent only a fraction of the possible "solution space" (Erb et al. 2017; Wurtzel et al. 2019). In fact, through different approaches, it has been shown that thousands of other CO$_2$-fixation pathways could theoretically exist, some of them with probably much higher efficiencies compared to the naturally evolved solutions. Yet, none of these pathways has apparently been realized by nature (Fig. 1; Bar-Even et al. 2010; Schwander et al. 2016; Trudeau et al. 2019; Löwe and Kremling 2021).

The expanding field of synthetic biology provides the exciting opportunity to realize these theoretical solutions from first principles. Unlike in evolution, where the emergence of novel CO$_2$-fixing pathways depends on the serendipitous recombination and integration of (existing) enzymes into preexisting metabolic networks, synthetic biology aims at constructing entirely new CO$_2$-converting pathways in a defined manner from the bottom up (Erb et al. 2017). This engineering approach of synthetic biology allows us to explore completely new-to-nature solutions that are of improved kinetic and thermodynamic efficiency, allowing higher fluxes (e.g., improved CO$_2$ capture) at lower energetic demand (e.g., ATP and NAD(P)H consumption) and also including completely new enzyme chemistries (Bar-Even et al. 2010; Schwander et al. 2016; Trudeau et al. 2019; Löwe and Kremling 2021). This radical approach of finding completely novel solutions (and thus a global optimum) separates synthetic biology from metabolic engineering approaches that typically aim at improving an existing solution (and thus finding a local optimum).

In summary, compared to classical and new breeding technologies, synthetic biology approaches have a disruptive potential allowing expansion of the existing metabolic solution space and opening the possibility to enhance plant productivity by developing a more efficient and productive photosynthesis compared to native plant metabolism (Erb et al. 2017; Wurtzel et al. 2019). Current efforts in the field have focused on two main challenges: (1) to boost natural photosynthesis by realizing new-to-nature photorespiration pathways (Trudeau et al. 2019; Scheffen et al. 2021), and (2) to entirely replace the CBB cycle of photosynthesis by synthetic CO$_2$-fixation pathways (Fig. 2; Schwander et al. 2016).

BOOSTING PHOTOSYNTHESIS THROUGH SYNTHETIC PHOTORESPIRATION PATHWAYS

As mentioned before, one of the biggest bottlenecks in photosynthesis is the wasteful process of photorespiration. This process uses more than a dozen enzymes, takes place in three different compartments, and consumes substantial amounts of energy (minimum of eight ATP and six reducing equivalents) to convert two molecules of 2-phosphoglycolate (2PG) back into one molecule or 3-phosphoglycerate (3PGA). Most importantly, however, up to 30% of the previously fixed carbon in photosynthesis is lost again to the environment through photorespiration, which actually limits carbon capture (Erb and Zarzycki 2016; Walker et al. 2016).

To increase photosynthetic yield, many previous (and current) efforts in plant engineering focus on improving the existing photorespiration pathway and/or realizing carbon concentrating mechanisms. These approaches include the implementation and transplantation of CO$_2$-transporters, carboxysomes, or metabolic carbon pumps, such as C4-metabolism into crops (Ort et al. 2015; Erb and Zarzycki 2016).

In contrast to these efforts, which treat the symptom rather than curing the root of the "disease," synthetic biology provides the opportunity to implement completely novel, synthetic photorespiration pathways with improved efficiency. In a recent study, several potential photorespiration pathways were designed that aim at converting 2PG, the product of photorespiration, back into a CBB or central carbon intermediate at minimal bioenergetic cost and without losing any CO$_2$ that had been fixed before (Trudeau et al. 2019). Following this initial theoretical phase, some of the most promising designs were explored recently both in vitro and in vivo.

Earlier efforts have already indicated that transplanting new metabolic traits into plants can indeed improve carbon capture and mitigate the effects of photorespiration (Kebeish et al. 2007; South et al. 2019). The most prominent example is an energetically more efficient photorespiration pathway that converts two molecules of glyoxylate into acetyl-CoA (and two CO$_2$).

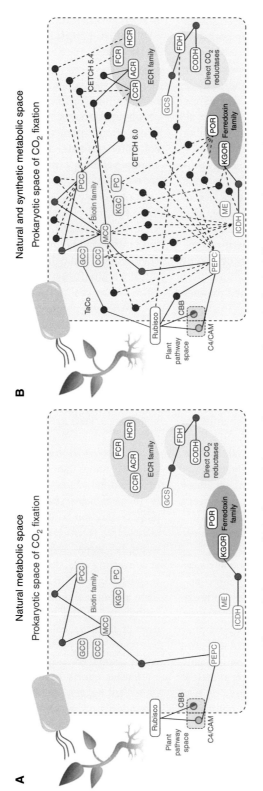

Figure 1. The metabolic space of CO_2-fixation. (*A*) Natural space: Nature has evolved eight different CO_2-fixation pathways (nodes) that are built around different classes of carboxylases (round boxes). (*B*) Natural and synthetic space: Shown are 28 additional new-to-nature pathways that have been designed, of which three have been realized in vitro (pink lines). Those 28 designs are covering only a fraction of the theoretical possible space. (Panel *B* is modified from Figure 1 in Wurtzel et al. 2019, with permission from the authors.)

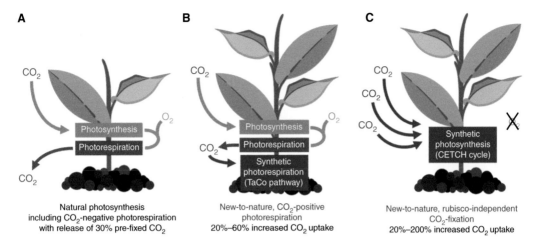

Figure 2. Two strategies to increase photosynthetic yield with synthetic biology. (*A*) Natural photosynthesis suffers from photorespiration caused by the oxygenation of rubisco that releases up to 30% of the pre-fixed carbon. (*B*) New-to-nature photorespiration pathways (such as the tatronyl-CoA [TaCo] pathway) capture CO$_2$ during photorespiration instead of releasing carbon, thus boosting photosynthetic yield. (*C*) New-to-nature synthetic photosynthetic pathways (such as the CETCH cycle) do not use rubisco as a CO$_2$-fixing enzyme, thus circumventing photorespiration. They are kinetically favored, as they use enzymes that fix CO$_2$ faster compared to rubisco and they require less energy, leading to an increased photosynthetic yield.

Transplanted into different crops, this pathway has resulted in biomass increases in the field (South et al. 2019). Although these pathways still are carbon-positive (i.e., release CO$_2$ during photorespiration), they demonstrate the proof-of-concept and lay the basis for more radical designs that are carbon-neutral or even carbon-negative (i.e., capture CO$_2$ during photorespiration instead of releasing it).

One example of a carbon-neutral pathway that was realized in a first effort is the β-hydroxyaspartate cycle (BHAC) (Schada von Borzyskowski et al. 2019; Roell et al. 2021). This pathway converts glyoxylate, a metabolite derived from plant photorespiration, into oxaloacetate in a highly efficient carbon-, nitrogen-, and energy-conserving manner. A functional BHAC was engineered in *Arabidopsis thaliana* peroxisomes to create a photorespiratory bypass that is independent of 3PGA or decarboxylation of photorespiratory precursors. Currently, efficient and diverse oxaloacetate conversion in *A. thaliana* still seems to mask the full potential of the BHAC (Roell et al. 2021). However, nitrogen conservation and accumulation of signature

metabolites demonstrate the proof-of-principle, which opens the door to further integration of this photorespiration-dependent synthetic carbon-concentrating mechanism in plants. In a complementary effort, the BHAC was recently also transferred into cyanobacteria, where it shows similar effects (Fig. 3; L Schada von Borzyskowksi, pers. comm.).

THE TaCo PATHWAY: A CO$_2$-FIXING PHOTORESPIRATION PATHWAY

Another particularly promising example is the TaCo pathway. The TaCo pathway was developed recently through computationally guided pathway design in combination with state-of-the art enzyme engineering, including microfluidics-based high-throughput screening (Trudeau et al. 2019; Scheffen et al. 2021). Notably, this new-to-nature photorespiration pathway allows additional capture of CO$_2$ instead of releasing it, which makes photorespiration for the first time a carbon-positive (i.e., CO$_2$-fixing) process, turning the "Achilles' heel" of photosynthesis into an asset. In other words, the TaCo pathway

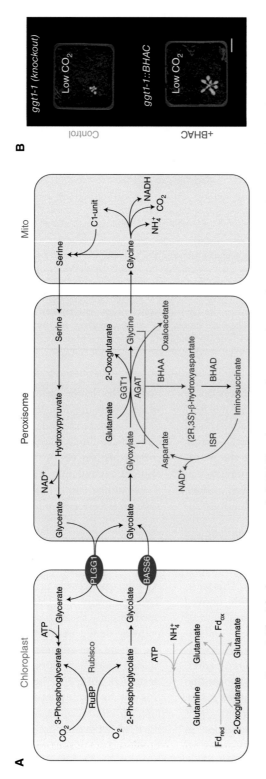

Figure 3. The β-hydroxyaspartate cycle (BHAC) pathway, a carbon-neutral CO_2-photorespiratory pathway. (*A*) Scheme of the integration of the BHAC pathway into *Arabidopsis thaliana* (photorespiration) metabolism. (*B*) Implementation of the BHAC in ggt1-1 mutant *A. thaliana* plants, mimicking a photorespiration phenotype at low (ambient) CO_2 concentrations, leads to increased growth yield compared to the control. (All panels are adapted from Roell et al. 2021, with permission from the authors who hold the copyright. Published by *PNAS* © 2021.)

would allow formation of biomass independent of whether rubisco has performed a carboxylation or oxygenation reaction, and thus overcome the most fundamental challenge that plants face during photosynthesis (Fig. 4).

The core reaction sequence of the TaCo pathway consists of three new-to-nature enzymes, including a novel CO$_2$-fixing enzyme, glycolyl-CoA carboxylase (GCC). GCC was developed in the scaffold of a biotin-dependent propionyl-CoA carboxylase through rational design, as well as random mutagenesis (Scheffen et al. 2021). During this process, the catalytic efficiency of GCC improved by three orders of magnitude to match the properties of natural CO$_2$-fixing enzymes, which has been key to realizing the TaCo pathway. Very recently, GCC was improved even further through machine learning approaches, giving rise to a highly efficient carboxylase (D Marchal, pers. comm.).

Modeling shows that the TaCo pathway dramatically reduces energetic demands of photorespiration by ∼30% (ATP) and 20% (NAD(P)H), respectively, while at the same time increasing carbon efficiency between 40% to 150%, independent of rubisco's oxygenation activity

(Fig. 5; Trudeau et al. 2019; Scheffen et al. 2021). This was further confirmed through biochemical experiments in which the TaCo was coupled with photorespiration and showed a 40% improvement of photosynthetic yield in vitro. Experiments to transfer the TaCo into plants have shown similar phenotypic effects.

REPLACING PHOTOSYNTHESIS THROUGH SYNTHETIC CO$_2$-FIXATION PATHWAYS

An even more radical approach is to entirely replace the CBB cycle by realizing a completely new-to-nature CO$_2$-fixation cycle that is not centered on rubisco (Erb and Zarzycki 2016; Schwander et al. 2016). As mentioned earlier, natural photosynthesis is limited by the inefficiency of its key enzyme, rubisco, which shows a very slow turnover rate of 5–10 CO$_2$ molecules per second and makes the enzyme a weak link in (any) CO$_2$-fixing metabolism.

Recently, a new class of CO$_2$-fixing enzymes, so-called enoyl-CoA carboxylases/reductases (ECRs), were discovered in alphaproteobacteria, where they serve in central acetate metabolism. These ECRs catalyze the NADPH-dependent

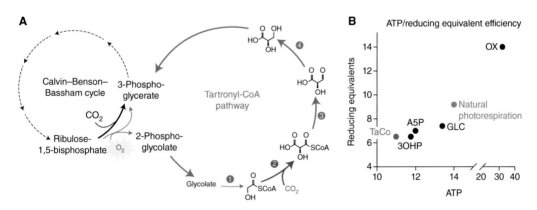

Figure 4. The tatronyl-CoA (TaCo) pathway, a carbon-positive photorespiration pathway. (*A*) Scheme of the TaCo pathway. The TaCo pathway requires four steps to convert glycolate to glycolyl-CoA (1, glycolyl CoA synthase), its carboxylation into tartronyl-CoA (2, glycolyl-CoA carboxylase), and its reduction into tatronic semialdeyhde and glycerate (3, bifunctional tatronyl-CoA reductase). The key step, the new-to-nature carboxylation reaction of glycolyl-CoA is highlighted in red. (*B*) Comparison of the TaCo pathway compared to other designs and the natural photorespiration pathway, respectively. (3OHP) 3-hydroxypropionate bypass, (A5P) arabinose-5-phosphate shunt, (GLC) glycerate bypass, (NPR) natural photorespiration, (OX) glycolate oxidation pathway. The TaCo pathway is by far the most energy and reducing-equivalent efficient solution. (Figure based on data in Scheffen et al. 2021.)

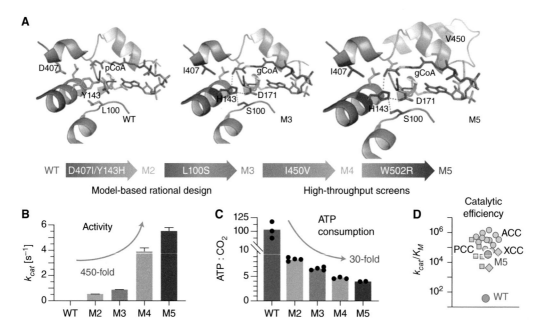

Figure 5. Engineering glycolyl-CoA carboxylase (GCC), the key enzyme of the tatronyl-CoA (TaCo) pathway. (*A*) GCC was developed in the scaffold of propionyl-CoA carboxylase (wild-type [WT]) through rational engineering (variants M2 and M3), as well as nontargeted directed evolution (M4 and M5). Final variant M5 contained five mutations. (*B*) Catalytic activity of the five different variants. While the WT showed almost no activity for gylcolyl-CoA carboxylation, variant M5 showed catalytic activities comparable to other CoA-carboxylases. (*C*) ATP consumption per CO_2 fixed for the five different variants. While the WT showed futile ATP hydrolysis, unfruitful ATP hydrolysis was strongly reduced in variant M5. (*D*) Catalytic efficiency shows that the M5 groups within other, naturally existing CoA-carboxylases. (All panels are modified from Figure 2 in Scheffen et al. 2021, with permission from Creative Commons Attribution 4.0 International License.)

reductive carboxylation of enoyl-CoA thioester into their corresponding alkyl-malonyl-CoA esters (Erb et al. 2007). ECRs are particularly interesting, because they capture CO_2 at up to 10-fold higher rates compared to rubisco (i.e., up to 100 CO_2 molecules per second). Moreover, they do not show any oxygenase side reaction (Erb et al. 2007, 2009; DeMirci et al. 2022). Overall, this makes ECRs an interesting alternative to rubisco-based CO_2-fixation.

Building on these ECRs, several new-to-nature CO_2-fixation pathways have been recently designed from the bottom up. A prime example of such efforts is the CETCH cycle. This cycle was designed around the carboxylation of crotonyl-CoA and acrylyl-CoA by ECRs (Schwander et al. 2016). The CETCH cycle is a network of 17 enzymes that is thermodynamically and kinetically favored compared to the Calvin cycle

of natural photosynthesis. The CETCH cycle requires 20% less energy and the carboxylation rates of crotonyl-CoA and acrylyl-CoA by ECR outcompete rubisco by almost one order of magnitude with respect to CO_2-fixation rates (Erb et al. 2007; Schwander et al. 2016; DeMirci et al. 2022).

The CETCH cycle was constructed by metabolic retrosynthesis and subsequently established with enzymes originating from nine different organisms of all three domains of life. Having established a first version of the cycle, the performance of the system was further optimized in several rounds including enzyme engineering and metabolic proofreading. An optimized version of the CETCH cycle (version 5.4) converts CO_2 into organic molecules faster than the natural CO_2-fixation pathway of photosynthesis and notably at 20% less energy per CO_2

fixed, reaching activities comparable to those of natural photosynthesis in vitro (Schwander et al. 2016).

To further optimize the CETCH cycle, a machine-learning-guided, high-throughput screening strategy, named METIS (named after the ancient goddess of wisdom and crafts Μῆτις, literally, "wise counsel") has been developed and applied recently (Pandi et al. 2022). This strategy used iterative design-build-learn cycles to search the combinatorial space of a complex in vitro system over several rounds for a (local) optimum. To find such optima, the workflow relies on automated experimentation for prototyping different combinations, which is followed by subsequent analysis and machine learning–guided prediction of an improved set of combinations (Fig. 6; Pandi et al. 2022).

The CETCH cycle features 26 different components, including 17 enzymes, several cofactors and salts, which span a theoretical space of ~10^{25} combinations in the chosen setup. Exploring this theoretical space over only eight rounds of active learning with ~1000 different variants improved CO_2-fixation productivity 10-fold compared to the initial described CETCH cycle version 5.4 (Schwander et al. 2016). These efforts resulted in the most efficient CO_2-fixation system described to date, demonstrating how machine learning can be combined with laboratory automation to systematically optimize synthetic biological networks in a data-driven fashion, before implementation in vivo.

With respect to the latter, it has already been demonstrated that the CETCH cycle can be coupled to chloroplast extracts forming an artificial chloroplast (Miller et al. 2020). Current efforts focus on the transplantation of the CETCH cycle into different microorganisms (Sánchez-Pascuala, unpubl.), which provides an excellent starting point for the further experimental evolution and implementation of the CETCH cycle in photosynthetic organisms and plants.

HOW DOES THE TECHNOLOGY SCALE?

Today, 450 Gt C are contained in plant biomass (60% above, 40% below ground) (Bar-On et al. 2018). However, the amount of plant biomass has declined about twofold since the beginning of human civilization, so that the actual global potential for photosynthesis is about 916 Gt C (Erb et al. 2018). This means that there is an additional pool of almost 470 Gt C for photosynthetic carbon capture. This additional pool exceeds the total amount of anthropogenic carbon released into the atmosphere over the last 150 yr (300 Gt C), thereby providing an enormous potential for photosynthetic carbon drawdown.

To avoid competition with food supplies, agriculturally used land cannot be simply repurposed for photosynthetic carbon drawdown. At the same time, given the need to maintain biodiversity and global ecosystems, the development of new arable land is limited. Consequently, more efficient carbon drawdown can only be realized when plant biomass productivity per land unit is increased.

Currently, about 18.6 million km^2 of cropland are globally available, which assimilate about 9.4 Gt y^{-1} (Jansson et al. 2021). How (much) can carbon assimilation be increased? Total photosynthetic yield (W_h) is defined as a fraction of the total incident solar energy during a growing season (S), light interception efficiency (ε_i), conversion efficiency (ε_c), and partitioning efficiency (η) through the following equation (Monteith 1977): $W_h = S \cdot \varepsilon_i \cdot \varepsilon_c \cdot \eta$.

While in modern crops, incident and partitioning efficiency have already been optimized, the current bottleneck in photosynthesis is conversion efficiency (ε_c), which strongly limits capture and conversion of carbon into biomass. Thus, any efforts to improve carbon fixation efficiency will directly increase photosynthetic yield (W_h). However, natural carbon fixation already runs at its limits (30%–50% of the total leaf proteome is rubisco and photorespiration results in carbon losses of up to 30%), which will require the introduction of new-to-nature solutions to overcome the limits of natural photosynthesis, as discussed above.

Based on photosynthetic models, it is predicted that a new-to-nature, CO_2-positive photorespiration through the TaCo mechanism will improve the carbon fixation rate between 20% and 60% (Trudeau et al. 2019), and photosynthetic carbon uptake in crops would increase by

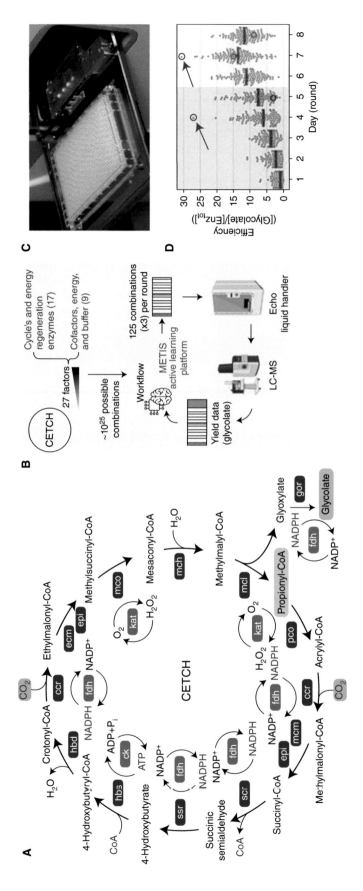

Figure 6. Design, realization, and optimization of the CETCH cycle. (*A*) Outline of the CETCH cycle. (*B*) Setup and workflow of the design-build-test cycle, including the METIS active learning platform. (*C*) Picture of laboratory automation with liquid-handling robots. (*D*) Improvement of productivity of the CETCH cycle over eight active learning cycles. Individual points represent different pathway variants. Pink arrows highlight highly favorable CETCH variants that are 10-fold improved in CO_2-fixation efficiency compared to initial version 5.4 and almost 60-fold compared to version 1.0 of the cycle. (Panels *A*, *B*, and *D* reprinted from Pandi et al. 2022 under the terms of a Creative Commons 4.0 International License.)

1.9 to 6.2 Gt C y^{-1}. For the CETCH cycle, thermodynamic calculations consistently predict a 20% increased efficiency (Schwander et al. 2016; Löwe and Kremling 2021), while kinetic benefits on the level of the carbon fixation step could be as high as 10-fold (Erb et al. 2007; DeMirci et al. 2022). Thus, estimates are that the carbon fixation rate could increase between 20% and 200% upon implementation of CETCH, which translates to an increased photosynthetic productivity in crops of 1.9 to 19 Gt C y^{-1}.

Although these calculations currently are based on rough estimates, they generally highlight the potential of synthetic biology to increase photosynthetic efficiency and have a real-world impact that could potentially scale.

WHAT IS REQUIRED TO MOVE FORWARD?

One challenge is our limited capability to engineer plant metabolism through the introduction of multigene constructs. Thus, shorter designs will very likely be implemented first (such as TaCo), before more radical designs (such as CETCH) can be realized. These efforts will be further supported and complemented by in vitro optimization of the different pathways, as well as the use of organisms with fast generation times as intermediate hosts to test and optimize the different designs in vivo (e.g., bacteria or photosynthetic microorganisms). Building up a pipeline that will allow swiftly moving from the theoretical design to the actual implementation in vivo could speed up the design and implementation cycle in the future.

In addition to the scientific challenges, there is also a great need to engage early on in a dialogue with society, stakeholders, and important policymakers about the risks and benefits of synthetic biology. Synthetic biology has met with mixed acceptance in different parts of the world. The unfolding of the climate crisis, however, has created a momentum for technology openness and a more balanced view on risks and benefits of organisms with new-to-nature CO$_2$-fixation pathways.

ACKNOWLEDGMENTS

This article has been made freely available online by generous financial support from the Bill and Melinda Gates Foundation. We particularly want to acknowledge the support and encouragement of Rodger Voorhies, president, Global Growth and Opportunity Fund.

REFERENCES

Bar-Even A, Noor E, Lewis NE, Milo R. 2010. Design and analysis of synthetic carbon fixation pathways. *Proc Natl Acad Sci* **107**: 8889–8894. doi:10.1073/pnas.0907176107

Bar-On YM, Phillips R, Milo R. 2018. The biomass distribution on Earth. *Proc Natl Acad Sci* **115**: 6506–6511. doi:10.1073/pnas.1711842115

Bassham JA, Benson AA, Kay LD, Harris AZ, Wilson AT, Calvin M. 1954. The path of carbon in photosynthesis. XXI: The cyclic regeneration of carbon dioxide acceptor. *J Am Chem Soc* **76**: 1760–1770. doi:10.1021/ja01636a012

Berg IA. 2011. Ecological aspects of the distribution of different autotrophic CO$_2$ fixation pathways. *App Environ Microbiol* **77**: 1925–1936. doi:10.1128/AEM.02473-10

Bierbaumer S, Nattermann M, Schulz L, Zschoche R, Erb TJ, Winkler CK, Tinzl M, Glueck SM. 2023. Enzymatic conversion of CO$_2$: from natural to artificial utilization. *Chem Rev* **123**: 5702–5754. doi:10.1021/acs.chemrev.2c00581

DeMirci H, Rao Y, Stoffel GM, Vögeli B, Schell K, Gomez A, Batyuk A, Gati C, Sierra RG, Hunter MS, et al. 2022. Intersubunit coupling enables fast CO$_2$-fixation by reductive carboxylases. *ACS Cent Sci* **8**: 1091–1101. doi:10.1021/acscentsci.2c00057

Erb TJ, Zarzycki J. 2016. Biochemical and synthetic biology approaches to improve photosynthetic CO$_2$-fixation. *Curr Opin Chem Biol* **34**: 72–79. doi:10.1016/j.cbpa.2016.06.026

Erb TJ, Berg IA, Brecht V, Müller M, Fuchs G, Alber BE. 2007. Synthesis of C$_5$-dicarboxylic acids from C$_2$-units involving crotonyl-CoA carboxylase/reductase: the ethylmalonyl-CoA pathway. *Proc Natl Acad Sci* **104**: 10631–10636. doi:10.1073/pnas.0702791104

Erb TJ, Brecht V, Fuchs G, Müller M, Alber BE. 2009. Carboxylation mechanism and stereochemistry of crotonyl-CoA carboxylase/reductase, a carboxylating enoyl-thioester reductase. *Proc Natl Acad Sci* **106**: 8871–8876. doi:10.1073/pnas.0903939106

Erb TJ, Jones PR, Bar-Even A. 2017. Synthetic metabolism: metabolic engineering meets enzyme design. *Curr Opin Chem Biol* **37**: 56–62. doi:10.1016/j.cbpa.2016.12.023

Erb KH, Kastner T, Plutzar C, Bais ALS, Carvalhais N, Fetzel T, Gingrich S, Haberl H, Lauk C, Niedertscheider M, et al. 2018. Unexpectedly large impact of forest management and grazing on global vegetation biomass. *Nature* **553**: 73–76. doi:10.1038/nature25138

Jansson C, Faiola C, Wingler A, Zhu XG, Kravchenko A, de Graff MA, Ogden AJ, Handakumbura PP, Werner C, Beckles DM. 2021. Crops for carbon farming. *Front Plant Sci* **12**: 636709. doi:10.3389/fpls.2021.636709

Kebeish R, Niessen M, Thiruveedhi K, Bari R, Hirsch HJ, Rosenkranz R, Stäbler N, Schönfeld B, Kreuzaler F, Peterhänsel C. 2007. Chloroplastic photorespiratory bypass increases photosynthesis and biomass production in

Arabidopsis thaliana. Nat Biotechnol **25**: 593–599. doi:10 .1038/nbt1299

Könneke M, Schubert DM, Brown PC, Hügler M, Standfest S, Schwander T, von Borzyskowski L S, Erb TJ, Stahl DA, Berg IA. 2014. Ammonia-oxidizing archaea use the most energy-efficient aerobic pathway for CO_2 fixation. *Proc Natl Acad Sci* **111**: 8239–8244. doi:10.1073/pnas .1402028111

Löwe H, Kremling A. 2021. In-depth computational analysis of natural and artificial carbon fixation pathways. *BioDesign Res* **2021**: 9898316.

Miller TE, Beneyton T, Schwander T, Diehl C, Girault M, McLean R, Chotel T, Claus P, Socorro Cortina N, Baret JC, et al. 2020. Bottom-up construction of a chloroplast mimic capable of light-driven synthetic CO_2 fixation. *Science* **368**: 649–654. doi:10.1126/science.aaz6802

Monteith JL. 1977. Climate and the efficiency of crop production in Britain. *Phil Trans R Soc Lond B Biol Sci* **281**: 277–294.

Ort DR, Merchant SS, Alric J, Blankenship RE, Bock R, Croce R, Hanson MR, Hibberd JM, Long SP, Moore TA, et al. 2015. Redesigning photosynthesis to sustainably meet global food and bioenergy demand. *Proc Natl Acad Sci* **112**: 8529–8536. doi:10.1073/pnas.1424031112

Pandi A, Diehl C, Yazdizadeh Kharrazi A, Scholz SA, Bobkova E, Faure L, Nattermann M, Adam D, Chapin N, Foroughijabbari Y, et al. 2022. A versatile active learning workflow for optimization of genetic and metabolic networks. *Nat Commun* **13**: 1–15. doi:10.1038/s41467-022-31245-z

Roell MS, Schada von Borzykowski L, Westhoff P, Plett A, Paczia N, Claus P, Urte S, Erb TJ, Weber APM. 2021. A synthetic C4 shuttle via the β-hydroxyaspartate cycle in C3 plants. *Proc Natl Acad Sci* **118**: e2022307118. doi:10 .1073/pnas.2022307118

Schada von Borzyskowski L, Severi F, Krüger K, Hermann L, Gilardet A, Sippel F, Pommerenke B, Claus P, Cortina NS, Glatter T, et al. 2019. Marine proteobacteria metabolize glycolate via the β-hydroxyaspartate cycle. *Nature* **575**: 500–504. doi:10.1038/s41586-019-1748-4

Scheffen M, Marchal DG, Beneyton T, Schuller SK, Klose M, Diehl C, Lehmann J, Pfister P, Carrillo M, He H, et al. 2021. A new-to-nature carboxylation module to improve natural and synthetic CO_2 fixation. *Nat Catal* **4**: 105–115. doi:10.1038/s41929-020-00557-y

Schulz L, Guo Z, Zarzycki J, Steinchen W, Schuller JM, Heimerl T, Prinz S, Mueller-Cajar O, Erb TJ, Hochberg GKA. 2022. Evolution of increased complexity and specificity at the dawn of form I Rubiscos. *Science* **378**: 155–160. doi:10 .1126/science.abq1416

Schwander T, Schada von Borzyskowski L, Burgener S, Cortina NS, Erb TJ. 2016. A synthetic pathway for the fixation of carbon dioxide in vitro. *Science* **354**: 900–904. doi:10 .1126/science.aah5237

South PF, Cavanagh AP, Liu HW, Ort DR. 2019. Synthetic glycolate metabolism pathways stimulate crop growth and productivity in the field. *Science* **363**: eaat9077. doi:10 .1126/science.aat9077

Trudeau DL, Edlich-Muth C, Zarzycki J, Scheffen M, Goldsmith M, Khersonsky O, Avizemer Z, Fleishman SJ, Cotton CAR, Erb TJ, et al. 2019. Design and in vitro realization of carbon-conserving photorespiration. *Proc Natl Acad Sci* **115**: E11455–E11464.

Walker BJ, VanLoocke A, Bernacchi CJ, Ort DR. 2016. The costs of photorespiration to food production now and in the future. *Annu Rev Plant Biol* **67**: 107–129. doi:10.1146/ annurev-arplant-043015-111709

World Population Prospects. 2019. World population prospects 2019: highlights. Department of Economic and Social Affairs, United Nations, New York. https://population.un.org/wpp/publications/files/wpp2019_highlights .pdf

Wurtzel ET, Vickers CE, Hanson AD, Millar AH, Cooper M, Voss-Fels KP, Nikel PL, Erb TJ. 2019. Revolutionizing agriculture with synthetic biology. *Nat Plants* **5**: 1207–1210. doi:10.1038/s41477-019-0539-0

Engineering Crassulacean Acid Metabolism in C$_3$ and C$_4$ Plants

Xiaohan Yang,[1,2] Yang Liu,[1] Guoliang Yuan,[3] David J. Weston,[1,2] and Gerald A. Tuskan[1,2]

[1]Biosciences Division; [2]The Center for Bioenergy Innovation, Oak Ridge National Laboratory, Oak Ridge, Tennessee 37831, USA; [3]Chemical and Biological Process Development Group, Pacific Northwest National Laboratory, Richland, Washington 99352, USA

Correspondence: yangx@ornl.gov

Carbon dioxide (CO$_2$) is a major greenhouse gas contributing to changing climatic conditions, which is a grand challenge affecting the security of food, energy, and environment. Photosynthesis plays the central role in plant-based CO$_2$ reduction. Plants performing CAM (crassulacean acid metabolism) photosynthesis have a much higher water use efficiency than those performing C$_3$ or C$_4$ photosynthesis. Therefore, there is a great potential for engineering CAM in C$_3$ or C$_4$ crops to enhance food/biomass production and carbon sequestration on arid, semiarid, abandoned, or marginal lands. Recent progresses in CAM plant genomics and evolution research, along with new advances in plant biotechnology, have provided a solid foundation for bioengineering to convert C$_3$/C$_4$ plants into CAM plants. Here, we first discuss the potential strategies for CAM engineering based on our current understanding of CAM evolution. Then we describe the technical approaches for engineering CAM in C$_3$ and C$_4$ plants, with a focus on an iterative four-step pipeline: (1) designing gene modules, (2) building the gene modules and transforming them into target plants, (3) testing the engineered plants through an integration of molecular biology, biochemistry, metabolism, and physiological approaches, and (4) learning to inform the next round of CAM engineering. Finally, we discuss the challenges and future opportunities for fully realizing the potential of CAM engineering.

The ever-increasing human population and changing climatic conditions together create a grand challenge that encompasses engineering secure supplies of food, feed, biomaterials, and fuels (Yang et al. 2015). One potential solution to this challenge is to enhance plant-based carbon dioxide (CO$_2$) reduction and ultimately the sustainable production of food and biomass on arid, semiarid, abandoned, or marginal lands through the engineering of crassulacean acid metabolism (CAM) (Yang et al. 2015, 2021). CAM is a special type of photosynthesis with a temporal separation of CO$_2$ fixation, in which (1) during the night, atmospheric CO$_2$ is captured through open stomata and fixed, via phosphoenolpyruvate carboxylase (PEPC), into malate, which is imported into the vacuole and converted to malic acid for storage, and (2) during the following daytime, the nocturnally accumulated malic acid is released from the vacuole and decarboxylated to release CO$_2$, which accumulates to high internal concentrations behind closed stomata and is re-fixed by ribulose-1,5-bisphosphate carboxylase/oxygenase (rubisco) in the Calvin–Benson–Bassham (CBB) cycle (Black and Osmond 2003; Yang et al. 2017; Winter and Smith 2022). CAM plants feature much higher water use efficiency (WUE) than plants performing either C_3 or C_4 photosynthesis because of the daytime closure of stomata (the pores on the leaf surface) for reducing water loss mediated by transpiration and the nighttime opening of stomata for CO_2 uptake in CAM plants (Borland et al. 2009). In arid conditions of the Sonoran Desert, greater CAM WUE results in increased biomass yield potential of 345% and 147% greater than that of C_3 and C_4 plants, respectively (Davis et al. 2014). Multiple previous reviews have discussed CAM engineering with a focus on the engineering principles and strategies of converting C_3 to CAM (Borland et al. 2014; Yang et al. 2015; Yuan et al. 2020; Schiller and Bräutigam 2021). Here, we provide an update on the CAM engineering strategies informed by the molecular evolution of CAM, describe how to engineer CAM in C_3 and C_4 plants using a design–build–test–learn (DBTL) approach, and discuss the challenges and opportunities for CAM engineering.

STRATEGIES FOR CAM ENGINEERING

Efforts for CAM engineering have been inspired by two aspects of CAM evolution: (1) convergent evolution of CAM from diverse C_3 plants (Yang et al. 2017, 2019; Heyduk et al. 2019), and (2) natural emergence of facultative CAM, which is environmentally triggered in a reversible manner in response to water-deficit or salinity stress in plants that, under well-watered conditions, perform C_3 photosynthesis (e.g., *Clusia pratensis*, *Mesembryanthemum crystallinum*) or C_4 photosynthesis (e.g., *Portulaca*) (Cushman et al. 2008; Winter and Holtum 2014; Winter 2019).

The core metabolic CAM machinery predates the divergence of C_3, CAM, and C_4 (Yin et al. 2018). Comparative genomics analysis reveals that the evolution from C_3 to CAM has mainly involved diel (diurnal cycle) rescheduling of gene expression relevant to stomatal movement, heat stress response, and carbohydrate me-

tabolism, along with sequence changes in a small number of genes (Yang et al. 2017). This is consistent with the hypothesis that the C_3-to-CAM transition involves the switch of malic acid accumulation from daytime to nighttime, which is driven by a fundamental metabolic reprogramming (Winter and Smith 2022). Indeed, C_3 and CAM plants share the same carboxylation and decarboxylation processes, but they differ in temporal patterns of activity: Carboxylation and decarboxylation occur during the nighttime and daytime, respectively, in CAM plants, whereas these two processes occur during the daytime and the nighttime, respectively, in C_3 plants (Santelia and Lawson 2016; Shameer et al. 2018). Therefore, engineering of CAM into C_3 plants should focus on temporal reprogramming (e.g., diel-cycle shift) of the expression of genes shared between the C_3 and CAM pathways (e.g., carboxylation, decarboxylation, stomatal movement), along with some other genetic modifications, including transferring a small number of CAM-specific genes, repressing or silencing endogenous C_3 plant genes not compatible with CAM, and engineering of leaf anatomical traits (e.g., succulence, cell size) (Yang et al. 2015; Yuan et al. 2020). The strategies for CAM engineering in C_3 plants are shown in Figure 1.

CAM and C_4 are similar in biochemistry including the primary CO_2 fixation mediated by PEPC and the secondary CO_2 fixation mediated by rubisco, but they feature temporal (i.e., day and night) and spatial (i.e., mesophyll and bundle sheath cells) separation of primary and secondary CO_2 fixation, respectively (Edwards 2019). It has been considered that CAM and C_4 plants have evolved independently from C_3 plants. Because CAM and C_4 photosynthesis are not found in identical cells of a plant, it has been hypothesized that there is incompatibility between these two photosynthetic pathways (Sage 2002). This is consistent with the lack of orthologous genes shared specifically by obligate CAM and C_4 species (Yin et al. 2018), suggesting that obligate CAM plants have not followed the evolutionary path from C_3 to C_4 to CAM. However, weak CAM can be induced in the C_4 plant *Portulaca oleracea* under drought stress (Winter and Holtum 2014), and it was recently reported that endogenous transcrip-

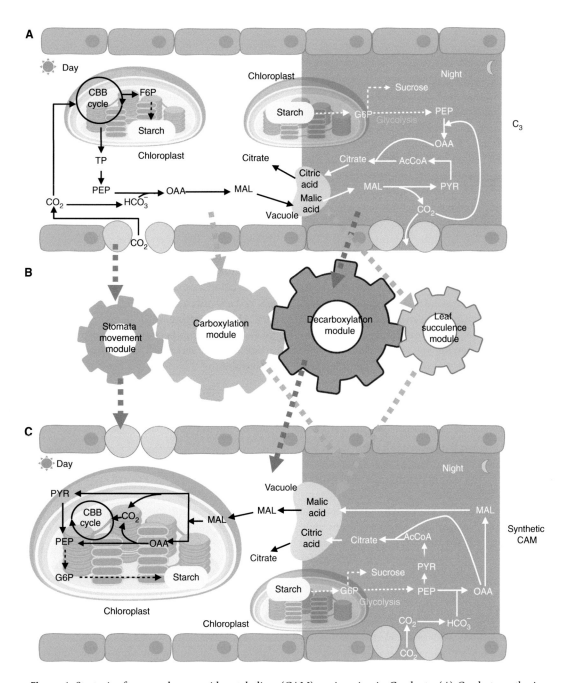

Figure 1. Strategies for crassulacean acid metabolism (CAM) engineering in C_3 plants. (A) C_3 photosynthesis pathway. (B) Genetic modifications for C_3-to-CAM transition. (C) Synthetic CAM pathway in C_3 plants. The dashed black (or white) lines indicate multistep metabolic processes. (AcCoA) Acetyl coenzyme A, (CBB) Calvin–Benson–Bassham, (F6P) fructose-6-phosphate, (G6P) glucose-6-phosphate, (MAL) malate, (OAA) oxaloacetic acid, (PEP) phosphoenolpyruvate, (PYR) pyruvate, (TP) triosephosphates. (Figure created from data in Noronha et al. 2018; Shameer et al. 2018; Rojas et al. 2019; Schiller and Bräutigam 2021; and Winter and Smith 2022.)

tion factor (TF) genes could play a critical role in the induction of CAM gene expression in drought-stressed leaves of *P. oleracea* (Ferrari et al. 2022), indicating that there is great potential for engineering CAM in C_4 plants through rewiring of diel gene expression. Because the carboxylation process, including the PEPC-mediated CO_2 fixation and the regeneration of phosphoenolpyruvate (PEP), operates in the mesophyll cells during the daytime in C_4 plants, it would be necessary to switch this process to the nighttime, along with one additional step (i.e., malic acid storage in the vacuole at night). Because decarboxylation operates during the daytime in both C_4 and CAM plants, there would be no need to change this process except adding one step for the efflux of malic acid from the vacuole during the daytime. Also, the changes in stomatal movement and leaf anatomical traits (e.g., succulence, cell size) for CAM engineering in C_4 plants can be the same as those for CAM engineering in C_3 plants. The strategies for CAM engineering in C_4 plants are shown in Figure 2.

Because seasonal drought stress has a broad impact on crop production, it would also be useful to engineer drought-inducible CAM or CAM-on-demand systems (Yuan et al. 2020). In natural facultative CAM plants, the transition from C_3/C_4 to CAM under drought stress is mediated by transcriptional regulation. For example, previous studies on drought-inducible gene expression have identified TF genes as candidate regulators of CAM induction in the facultative CAM species *M. crystallinum* (in TF families AP2/ERF, MYB, WRKY, NAC, NF-Y, and bZIP) (Amin et al. 2019) and *P. oleracea* (including TF genes *HB7*, *NFYA7*, *NFYC9*, *TT8*, and *ARR12*) (Ferrari et al. 2022). However, the roles of these candidate TF genes in drought-inducible CAM have not been experimentally validated yet; therefore, strategies for the engineering of synthetic facultative CAM based on TF engineering are currently not feasible. A more realistic strategy for creating facultative CAM would be to add the components for drought-responsive gene expression on top of the synthetic obligate CAM engineered in C_3 or C_4 plants (see more details in the Designing Synthetic CAM Pathways section).

Although the strategies for CAM engineering in C_3 or C_4 plants seem relatively straightforward, it may require multiple cycles of DBTL to optimize the design of synthetic CAM pathways in C_3 or C_4 plants (Yang et al. 2020). In the following, we will discuss how to implement the CAM engineering strategies using a DBTL approach.

DESIGNING SYNTHETIC CAM PATHWAYS

Modularity is one important principle of plant biosystems design (Yang et al. 2020). Based on the strategies described above, the design of CAM engineering in C_3 or C_4 plants can be divided into four modules: carboxylation module, decarboxylation module, stomatal movement module, and leaf succulence module. These modules for CAM engineering will be discussed separately for C_3 and C_4 plants because of a major difference in biochemistry between C_3 and C_4 photosynthetic pathways.

Synthetic Obligate CAM in C_3 Plants

Carboxylation Module

The carboxylation module for CAM engineering in C_3 plants can be designed in two different ways. The first is to express all the carboxylation genes derived from CAM plants, including genes for converting CO_2 and PEP to malate, importing malate from the cytoplasm into the vacuole, and recovering PEP from stored carbohydrates (e.g., starch) during the nighttime while repressing the expression of endogenous genes orthologous to the CAM carboxylation genes during the daytime using CRISPR interference (CRISPRi) (Fig. 3A). The second approach is to activate the expression of endogenous genes orthologous to the CAM carboxylation genes during the nighttime using CRISPR activation (CRISPRa) while repressing the expression of these endogenous genes during the daytime using RNA interference (RNAi) (Fig. 3B).

Decarboxylation Module

The decarboxylation module for CAM engineering in C_3 plants can also be designed in two dif-

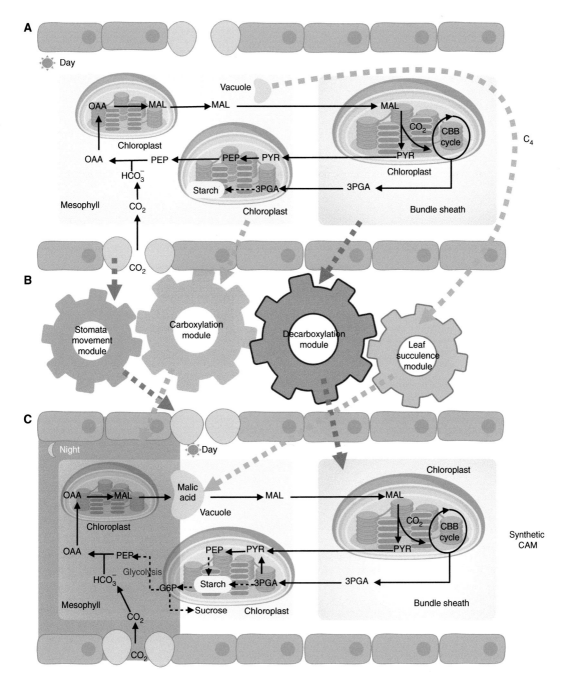

Figure 2. Strategies for crassulacean acid metabolism (CAM) engineering in C$_4$ plants. (*A*) C$_4$ photosynthesis pathway. (*B*) Genetic modifications for C$_4$-to-CAM transition. (*C*) Synthetic CAM pathway in C$_4$ plants. The dashed black lines indicate multistep metabolic processes. (CBB) Calvin–Benson–Bassham, (G6P) glucose-6-phosphate, (MAL) malate, (OAA) oxaloacetic acid, (3PGA) 3-phosphoglycerate, (PEP) phosphoenolpyruvate, (PYR) pyruvate. (Figure created from data in Friso et al. 2010; Furbank and Kelly 2021; Schiller and Bräutigam 2021; Gilman et al. 2022; and Reyna-Llorens and Aubry 2022.)

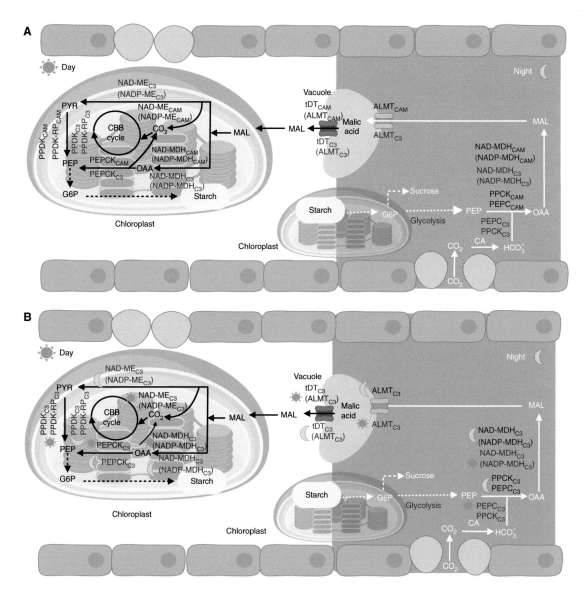

Figure 3. Design of crassulacean acid metabolism (CAM) carboxylation and CAM decarboxylation modules in C_3 host plants. (*A*) Option 1—ectopic expression of CAM pathway genes in C_3 plants while repressing the expression of endogenous genes in C_3 plants, which are orthologs of the genes involved in the CAM pathway in CAM plants. (*B*) Option 2—rewiring the expression of endogenous genes in C_3 plants, which are orthologs of the genes involved in the CAM pathway in CAM plants. Protein names in red and purple fonts indicate the activation (or overexpression) and repression, respectively, of corresponding genes. The glycolysis involves multiple enzymes including HK, PFK1, PFKFB, PK, LDH, aldolase, GAPDH, PGAM, and enolase (Bian et al. 2022). (ALMT) Aluminum-activated malate transporter, (CA) carbonic anhydrase, (CBB) Calvin–Benson–Bassham, (GAPDH) glyceraldehyde-3-phosphate dehydrogenase, (G6P) glucose-6-phosphate, (HK) hexokinase, (LDH) lactate dehydrogenase, (MAL) malate, (NAD-MDH) nicotinamide adenine dinucleotide malate dehydrogenase, (NAD-ME) NAD-dependent malic enzyme, (NADP-MDH) nicotinamide ADP malate dehydrogenase, (NADP-ME) NADP-dependent malic enzyme, (OAA) oxaloacetic acid, (PEP) phosphoenolpyruvate, (PEPC) PEP carboxylase, (PEPCK) PEP carboxykinase, (PFK1) phosphofructokinase 1, (PFKFB) 6-phosphofructo-2-kinase/fructose-2,6-biphosphatase, (PGAM) phosphoglycerate mutase, (PK) pyruvate kinase, (PPCK) PEPC kinase, (PPDK) PYR orthophosphate dikinase, (PPDK-RP) PPDK-regulatory protein, (PYR) pyruvate, (tDT) tonoplast dicarboxylate transporter, (TP) triosephosphates. The dashed lines indicate multistep metabolic processes.

 Cite this article as *Cold Spring Harb Perspect Biol* doi: 10.1101/cshperspect.a041674

ferent ways. The first is to express all the decarboxylation genes derived from CAM plants, including genes for exporting malate from the vacuole to the cytoplasm and releasing CO$_2$ from malate during the daytime while repressing the expression of endogenous genes orthologous to the CAM decarboxylation genes during the nighttime using CRISPRi (Fig. 3A). The second is to activate the expression of endogenous genes orthologous to the CAM decarboxylation genes during the daytime using CRISPRa while repressing the expression of these endogenous genes during the nighttime using RNAi (Fig. 3B).

Stomatal Movement Module

The design of the stomatal movement module can be informed by the molecular mechanism of stomatal movement in response to light and darkness in C$_3$ or C$_4$ plants. Stomatal opening and closing are driven by the reversible swelling and shrinking of guard cells, respectively, which are controlled by the ion channels and ion transporters in the plasma membrane and vacuolar membrane of guard cells (Pandey et al. 2007). Specifically, the hydroactive stomatal closure involves (1) the efflux of anions (e.g., Cl$^-$ and NO$_3^-$), predominantly via slow (S)-type channels from the SLOW ANION CHANNEL1 (SLAC1)/SLAC1 HOMOLOG (SLAH) family (e.g., SLAC1 and its homolog SLAH3), (2) extrusion of organic acids (e.g., malate) mediated by rapid (R)-type channels from the aluminum-activated malate transporter (ALMT) family such as ALMT12, which is also referred to as QUAC1 (quick-activating anion channel 1), and (3) release of K$^+$ by outward-rectifying Shaker K$^+$ channels, such as GORK (Sussmilch et al. 2019; Kashtoh and Baek 2021). Recently, it was reported that expressing the algae *Guillardia theta* anion channelrhodopsin 1 (*Gt*ACR1) in tobacco plants could provoke stomatal closure at conditions that induce stomatal opening in wild-type plants (Huang et al. 2021a; Jones et al. 2022). On the other hand, the hydroactive stomatal opening is mainly driven by the uptake of K$^+$ via inward-rectifying Shaker channels, such as KAT1, KAT2, AKT1, AKT2, and *At*KC1 (Sussmilch et al. 2019; Kashtoh and

Baek 2021). Also, TFs *At*Myb60 and *At*Myb61 are involved in light-induced stomatal opening and dark-induced stomatal closing, respectively, in C$_3$ plants (Cominelli et al. 2010).

The stomatal movement module for CAM engineering in C$_3$ plants can be further divided into two submodules: the stomatal closing submodule and the stomatal opening submodule. The stomatal closing submodule activates the endogenous genes responsible for stomatal closing while repressing the expression of the endogenous genes responsible for stomatal opening during the daytime (Fig. 4A). Also, overexpression of the *Gt*ACR1 from algae (Huang et al. 2021a; Jones et al. 2022) can be included in the design of the stomatal closing submodule. The stomatal opening submodule activates the endogenous genes responsible for stomatal opening while repressing the expression of endogenous genes responsible for stomatal closing during the nighttime (Fig. 4B).

Leaf Succulence Module

It has been reported that overexpression of a codon-optimized *Vitis vinifera* helix–loop–helix transcription factor (*Vv*CEB1) in the C$_3$ plant *Arabidopsis thaliana* increased leaf cell size and tissue succulence (i.e., ratio of tissue water/leaf area or dry mass) (Lim et al. 2018, 2020). Also, it has been demonstrated that overexpression of a tobacco tonoplast intrinsic protein gene (*Nt*TIP1;1) in tobacco BY-2 cells (*Nicotiana tobacum* L. cv. Bright Yellow 2) increased vacuolar volume and cell size (Okubo-Kurihara et al. 2009). Overexpression of these genes (i.e., *Vv*CEB1, *Nt*TIP1;1) under constitutive promoters should increase leaf succulence in C$_3$ plants.

Synthetic Obligate CAM in C$_4$ Plants

Carboxylation Module

Because C$_4$ and CAM are nearly identical in biochemistry, the design of the carboxylation module for CAM engineering in C$_4$ plants should be simpler than that in C$_3$ plants. One option for engineering the carboxylation module in C$_4$ plants is to (1) activate the expression of endogenous genes for converting CO$_2$ and PEP to malate

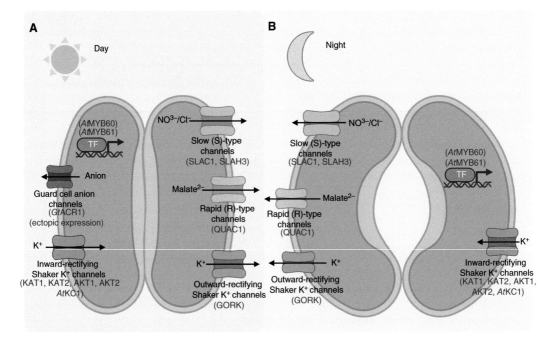

Figure 4. Design of crassulacean acid metabolism (CAM) stomatal movement module in C$_4$ host plants. (*A*) stomatal closing submodule. (*B*) stomatal opening submodule. Protein names in red and purple fonts indicate the activation (or overexpression) and repression, respectively, of corresponding genes. (AKT) *Arabidopsis* K$^+$ transporter, (*At*KC1) *Arabidopsis* Shaker-like K$^+$ channel 1, (*At*MYB60) *Arabidopsis* MYB domain protein 60, (*At*MYB61) *Arabidopsis* MYB domain protein 61, (GORK) guard cell outward-rectifying K$^+$ channel, (*Gt*ACR1) *Guillardia theta* anion channelrhodopsin 1, (KAT) K$^+$ channel in *Arabidopsis thaliana*, (SLAC1) slow anion channel 1, (SLAH3) SLAC1 homolog 3, (QUAC1) quick anion channel 1, (TF) transcription factor. (Figure created from data in Cominelli et al. 2010; Kollist et al. 2014; and Kashtoh and Baek 2021.)

using CRISPRa and overexpress the genes for importing malate from the cytoplasm into the vacuole and recovering PEP from stored carbohydrates (e.g., starch) during the nighttime, and (2) repress the expression of these genes during the daytime using RNAi (Fig. 5).

Decarboxylation Module

The decarboxylation module for CAM engineering in C$_4$ plants is straightforward because the endogenous genes for releasing CO$_2$ from malate are active during the daytime. One option for engineering the decarboxylation module in C$_4$ plants is to overexpress the gene for exporting malate from the vacuole to the cytoplasm during the daytime while repressing the expression of the same gene during the nighttime using CRISPRi (Fig. 5).

Stomatal Movement Module

The design of the stomatal movement module in C$_4$ plants can follow the same design in C$_3$ plants as described above (Fig. 4).

Leaf Succulence Module

The design of a leaf succulence module in C$_4$ plants can also follow the same design in C$_3$ plants as described above.

Synthetic Facultative CAM in C$_3$ or C$_4$ Plants

As mentioned in the Strategies for CAM Engineering section, a potential strategy for creating synthetic inducible CAM is to modify the design of synthetic obligate CAM by adding the components for drought-responsive gene expression. To

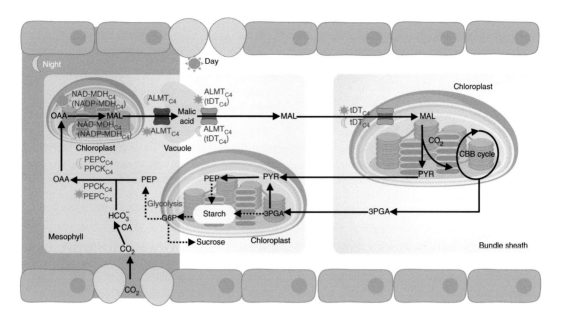

Figure 5. Design of crassulacean acid metabolism (CAM) carboxylation and CAM decarboxylation modules in C_4 host plants. Protein names in red and purple fonts indicate the activation (or overexpression) and repression, respectively, of corresponding genes. The glycolysis involves multiple enzymes including HK, PFK1, PFKFB, PK, LDH, aldolase, GAPDH, PGAM, and enolase (Bian et al. 2022). (ALMT) aluminum-activated malate transporter, (CA) carbonic anhydrase, (CBB) Calvin–Benson–Bassham, (GAPDH) glyceraldehyde-3-phosphate dehydrogenase, (G6P) glucose-6-phosphate, (HK) hexokinase, (LDH) lactate dehydrogenase, (MAL) malate, (NAD-MDH) nicotinamide adenine dinucleotide malate dehydrogenase, (NADP-MDH) nicotinamide ADP malate dehydrogenase, (OAA) oxaloacetic acid, (PEP) phosphoenolpyruvate, (PEPC) PEP carboxylase, (PFK1) phosphofructokinase 1, (PFKFB) 6-phosphofructo-2-kinase/fructose-2,6-biphosphatase, (3PGA) 3-phosphoglycerate, (PGAM) phosphoglycerate mutase, (PK) pyruvate kinase, (PPCK) PEPC kinase, (PYR) pyruvate, (tDT) tonoplast dicarboxylate transporter. The dashed lines indicate multistep metabolic processes.

implement this strategy, a "Boolean logic AND gate" can be designed to integrate two types of signal inputs—that is, the drought stress signal and the light/dark signal (Yuan et al. 2020). For example, a drought-inducible temporal activation system can be designed for CRISPRa-mediated regulation of genes in the carboxylation, decarboxylation, and stomatal movement modules by combining drought-inducible expression of the dCas9 gene and light/dark-inducible expression of guide RNA (gRNA) transcription units.

BUILDING SYNTHETIC CAM PATHWAYS

The "Build" stage of a DBTL cycle for CAM engineering involves two steps: (1) creating the gene constructs corresponding to the gene modules described in the Designing Synthetic CAM Path-

ways section, and (2) transforming the gene constructs into target C_3 or C_4 species.

Creating the Gene Constructs

The genes in each module can be cloned into a single plasmid vector, with each gene driven via a constitutive promoter or light/dark-inducible promoter depending on the temporal design of gene expression described in the Designing Synthetic CAM Pathways section. A recent quantitative analysis of plant-based promoters by Tian et al. (2022) provides useful information for the selection of different constitutive promoters to achieve optimal control of gene expression in plants. Based on this approach, multigene constructs for CAM engineering, designed to avoid repeated use of the same promoter, which may cause gene silencing in the transgenic plants, can be achieved. In addition, the selection of transcription terminators, paired with constitutive promoters, will be crucial for optimizing transgene expression in plants (Wang et al. 2020). As such, different combinations of constitutive promoters and terminators need to be tested for CAM engineering to optimize the metabolic flux.

Regarding the gene constructs for daytime-specific modules (e.g., decarboxylation module, stomatal closure submodule), promoters responsive to ambient white light are needed to enable daytime-specific gene expression. However, previously identified light-inducible promoters are largely responsive to blue light, red light, or green light because ambient white light leads to undesired system activation (Ochoa-Fernandez et al. 2020; Omelina et al. 2022). Recently, a white light–inducible bidirectional promoter (P_{Bn265}) was identified in *Brassica napus* (Zhu et al. 2022). Additional white light–inducible promoters will be needed to facilitate future engineering of daytime-specific modules.

To create the gene constructs for night-specific modules (e.g., carboxylation module, stomatal opening submodule), dark-inducible promoters are needed to enable nighttime-specific gene expression. Known dark-inducible promoters include *sen1*, *din6*, and *din10* (Fujiki et al. 2001). Additional dark-inducible promoters will need to be identified through transcriptome-sequencing (RNA-seq) analysis of leaf gene expression induced by dark treatment.

Besides the direct use of promoters to drive CAM-related genes for CAM engineering, CRISPRa tools, such as CRISPR–Act3.0 (Pan et al. 2021b), can be used to activate the expression of endogenous genes in the C_3 or C_4 plants, with gRNAs targeting the promoter regions of endogenous genes. Furthermore, CRISPRi (Pan et al. 2021a) can be used to repress the expression of endogenous genes, which show inverted day/night expression pattern in comparison with their orthologous genes in CAM plants. gRNA arrays with multiple gRNAs driven by a single promoter (Hassan et al. 2021) can be integrated with CRISPRa or CRISPRi to regulate the expression of multiple target genes. Recently, a new platform, called PARA, was developed for the rapid construction of gRNA arrays (Yuan et al. 2022b). This new cloning system can be used to create multiplex CRISPRa or CRISPRi constructs for CAM engineering.

Transforming the Gene Constructs into Target C_3 or C_4 Species

Using gene constructs for CAM engineering for the transformation of C_3 or C_4 plants via *Agrobacterium*-mediated plant transformation systems, which have been established for most of the crop plant species, such as rice (Jain et al. 2022), sorghum (Aregawi et al. 2022; Nelson-Vasilchik et al. 2022), maize (Anand et al. 2018; Peterson et al. 2021), pennycress (McGinn et al. 2019), switchgrass (Prías-Blanco et al. 2022; Xu et al. 2022), and poplar (Song et al. 2019; Yuan et al. 2021; Wen et al. 2022), is a promising and credible approach. Recently, a UV-visible reporter, called eYGFPuv, was used to achieve stable plant transformation, allowing for early and efficient selection of transformants as well as noninvasive visualization and easy tracking of transgenic events after transformation (Yuan et al. 2021). This eYGFPuv-based selection and visualization method can be used to facilitate the transformation of CAM gene modules into C_3 or C_4 plants.

TESTING SYNTHETIC CAM PATHWAYS

The transgenic plants created from CAM engineering should be tested across all levels of molecular biology, biochemistry, metabolism, and physiology.

Testing Engineered Plants at the Level of Molecular Biology

Transcriptome sequencing (RNA-seq) has been widely used for profiling the diel changes in transcript abundance in obligate CAM species, such as *Ananas comosus* (Ming et al. 2015), *Agave americana* (Abraham et al. 2016), and *Kalanchoë fedtschenkoi* (Yang et al. 2017), as well as inducible CAM species, such as *M. crystallinum* (Cushman et al. 2008) and *P. oleracea* (Ferrari et al. 2022). Similar RNA-seq approaches should be used to test whether the expression profiles of CAM-related genes are changed by CAM engineering

Cite this article as *Cold Spring Harb Perspect Biol* doi: 10.1101/cshperspect.a041674

in the transgenic plants. Single-cell RNA-seq (scRNA-seq), which enables the comprehensive analysis of gene expression in both common and rare cell types in plants (Bawa et al. 2022), can be used for studying gene expression in different cell types relevant to CAM, such as mesophyll cells and guard cells.

Because the temporal dynamics of protein abundance is not always consistent with that of the corresponding transcript abundance during a diel cycle, as demonstrated in the obligate CAM plant *A. americana* (Abraham et al. 2016), it will be necessary to perform proteomics analysis of the transgenic plants generated from CAM engineering.

Testing Engineered Plants at the Level of Biochemistry

For testing the impact of CAM engineering, it is critical to determine the kinetic parameters of key enzymes and the activities of transporters in the synthetic CAM pathways, such as PEPC and malic acid transporters. The kinetic properties of enzymes, such as Michaelis constant (K_m), maximal velocity (V_{max}), and turnover number (k_{cat}) can be determined for their substrates using spectrophotometry (Bläsing et al. 2000; Bertea et al. 2001). In addition, the activities of transporters can be measured using electrophysiological approaches (Hafke et al. 2003; Kovermann et al. 2007; Li et al. 2022). These data would be used for modeling analysis in the "Learn" stage of the DBTL cycle.

Testing Engineered Plants at the Level of Metabolism

For testing the metabolic changes, metabolites related to the CAM pathway can be analyzed using a metabolomics approach (Abraham et al. 2016). Furthermore, ^{13}C metabolic flux analysis (^{13}C-MFA) should be used for determining intracellular fluxes (Long and Antoniewicz 2019).

Testing Engineered Plants at the Level of Physiology

In CAM plants, stomata open and CO$_2$ is metabolized into malic acid and stored within vacuoles at night. During the day when stomata remain closed, the malic acid derivative malate is excreted from the vacuoles and transported to chloroplasts where it is cleaved into pyruvate and CO$_2$. The CO$_2$ is then available for rubisco-mediated photosynthesis. This concomitant change in leaf tissue acidity throughout the day and night cycle is a hallmark feature of CAM and easily detectable with acid titration (Zhang et al. 2020) or metabolomic approaches (Abraham et al. 2016). Another key physiological feature of CAM is diel changes in stomatal conductance and photosynthesis. As noted above, in CAM plants stomata open at night thereby increasing water and CO$_2$ conductance relative to daytime values. By using readily available infrared gas analyzers, the rates of both CO$_2$ and water vapor can be measured. When performed across a diel day/night cycle, measurements of pH, CO$_2$, and water vapor flux can provide key physiology data for determining and evaluating CAM engineering efforts. The level of CAM physiology expression can be indicated by the relative proportion of CO$_2$ uptake in the dark over a 24-h period, with nighttime CO$_2$ fixation contributing >70% and <1/3 to total carbon gain considered as strong and weak CAM, respectively (Nelson and Sage 2008).

LEARNING FROM ENGINEERED CAM PATHWAYS

The last stage of a DBTL cycle for CAM engineering is to learn from the experimental data generated from the "Test" stage through metabolic modeling and system dynamics (SD) modeling.

Metabolic Modeling

A genome-scale metabolic model (GSM), which is an in silico metabolic model comprising hundreds or thousands of chemical reactions in the metabolic inventory of an organism, is useful for mapping carbon flux through metabolic pathways (Simons et al. 2014; Beilsmith et al. 2022). A GSM can be integrated with flux balance analysis to predict cellular metabolic flux distributions (Simons et al. 2014). GSMs have been constructed in multiple plant species, including C$_3$ species (e.g., *A. thaliana*, *B. napus*, *Oryza sativa*) and C$_4$

species (e.g., *Setaria italica*, *Sorghum bicolor*, *Zea mays*) (Clark et al. 2020). Fluxomic data generated from ^{13}C-MFA analysis have been successfully used to develop a genome-scale kinetic model of *Escherichia coli* (Khodayari and Maranas 2016). A similar approach should be applied to develop genome-scale kinetic models of the genetically modified plants generated through CAM engineering. Specifically, the GSMs of target C_3 or C_4 plants can be modified to include the synthetic CAM pathways described in the Designing Synthetic CAM Pathways section. Then, genome-scale kinetic models can be generated through an integration of the modified GSMs and fluxomic data generated from ^{13}C-MFA analysis. Recently, a two-cell, two-phase flux balance model was built for modeling the efficiency of possible C_4 and CAM configurations in the facultative CAM species *P. oleracea*, which includes all major metabolic enzymes and reactions highly conserved across a wide array of plant genomes (Moreno-Villena et al. 2022). An analogous approach can be adapted for the metabolic modeling of synthetic CAM in C_4 plants. Also, a large-scale constraint-based modeling of metabolism and energetics in C_3 and CAM plants suggests that there are incremental changes in metabolic fluxes during the evolution of CAM from C_3 (Tay et al. 2021). A similar approach can be used for modeling the C_3-to-CAM transition in the transgenic plants.

System Dynamics Modeling

In addition to GSMs, SD models have been used to model CAM expression (i.e., diel pattern of gas exchange and malic acid accumulation) from measured biochemical constants (e.g., enzyme kinetic parameters) and physiological constants (e.g., stomatal and mesophyll conductance, rate of malic acid decarboxylation, rate of mitochondrial respiration, vacuole capacity) (Owen and Griffiths 2013). The SD modeling approach should be extended to include the spatial and temporal concentration of RNAs and proteins (Borland and Yang 2013). The causal relationship between the input (e.g., biochemical and physiological constants, RNA, and protein expression) and output (CAM expression) can be established through SD modeling analysis of data generated in the "Test" stage

(see Testing Synthetic CAM Pathways section) of the DBTL cycle.

CHALLENGES AND OPPORTUNITIES

Although significant progress has been made in the identification of the design principles and building blocks necessary for CAM engineering, there are still considerable knowledge gaps and technical challenges that need to be addressed before the potential of CAM engineering in C_3 or C_4 crops can be realized. Some of the knowledge gaps include the following.

1. The minimum number of genes that are needed for CAM engineering has not been determined yet. However, it was recently reported that overexpressing a CAM-specific *A. americana* PEPC gene (*AaPEPC1*) in tobacco upregulated multiple orthologs of CAM pathway genes, such as the genes encoding carbonic anhydrase (CA), malate dehydrogenase (MDH), aluminum-activated malate transporter (ALMT), tonoplast dicarboxylate transporter (tDT), and malic enzyme (ME) (Liu et al. 2021). Based on this information, it appears that CA, MDH, and ALMT can be omitted from the carboxylation module for CAM engineering. It is possible that there are key TFs acting as master regulators of CAM pathway genes. Thus, the number of genes for CAM engineering could be significantly reduced by shifting the focus from rewiring the expression of the genes encoding the CAM pathway enzymes and transporters to the manipulation of these master transcriptional regulators.

2. There is a lack of understanding of the evolutionary conservation and diversification of enzymes and transporters among C_3, C_4, and CAM plants. There is a need for systemically comparing the activities of carboxylation and decarboxylation enzymes among C_3, C_4, and CAM plants to identify the best isoform of each enzyme to maximize metabolic flux in synthetic CAM pathways.

3. The vacuolar transporters for malic acid import and export have not been fully characterized (Huang et al. 2021b). Further inves-

tigation will be needed to gain a deeper understanding of the molecular mechanism underlying malate transport in relation to CAM engineering.

4. There is a lack of information about the properties (kinetic parameters) of enzymes involved in the synthetic CAM pathway. Future efforts will be needed to generate quantitative information about the activity of each enzyme to be used for CAM engineering. This type of quantitative information will be very important for optimizing the design of the synthetic CAM pathway.

The technical challenges for CAM engineering include the following.

1. CAM engineering, based on the above design of the synthetic CAM pathway, will require transformations of 20 to 30 or even more genes in the C$_3$ or C$_4$ plants. This represents a current challenge based on the widely used *Agrobacterium*-mediated transformation. This challenge could be mitigated by using a newly developed dual-vector system based on a split selectable marker (Yuan et al. 2022a), which allows for transforming twice the number of genes as the traditional single-vector systems could handle.

2. Currently, it is challenging to rewire the diel expression patterns of a relatively large number of genes for C$_3$/C$_4$-to-CAM transition. To address this challenge, new capabilities based on homologous recombination-mediated precision genome editing (Hassan et al. 2022) need to be developed for promoter engineering.

3. Finally, CAM engineering may generate unintended impacts on plant metabolisms. To address this problem, the expression of synthetic CAM pathway genes needs to be tissue-/cell-type-specific, such as mesophyll cell–specific expression of carboxylation genes and guard cell–specific expression of stomatal movement genes. It is likely that cell-type-specific genes can be identified through scRNA-seq analysis of leaf gene expression (Liu et al. 2022; Xia et al. 2022).

With the rapid advancement of new plant functional genomics and synthetic biology technologies, the knowledge gaps and technical challenges related to engineering synthetic CAM in C$_3$ or C$_4$ crops are not insurmountable, suggesting that engineered C$_3$ or C$_4$ plants viably expressing CAM machinery could be feasible soon. It is expected that CAM engineering would be able to make a significant contribution to climate change mitigation through the enhancement of plant-based CO$_2$ capture and storage. Approximately 40% of the world's land surface is arid or semiarid (Jia et al. 2021). Engineering of synthetic CAM pathways would enable sustainable production of crops on most of these arid and semiarid lands, with great potential for increasing plant-based carbon sequestration. Some candidate target plants for CAM engineering include food crops (e.g., rice, sorghum, maize, soybean), bioenergy crops (e.g., poplar, pennycress, switchgrass), horticultural crops (e.g., apples, peaches, pecans), vegetable crops (tomatoes, onions, cauliflower), and landscape crops (e.g., turfgrasses, woody ornamental plants).

COMPETING INTEREST STATEMENT

This manuscript has been authored by UT-Battelle, LLC under Contract No. DE-AC05-00OR22725 with the U.S. Department of Energy. The United States Government retains and the publisher, by accepting the article for publication, acknowledges that the United States Government retains a non-exclusive, paid-up, irrevocable, worldwide license to publish or reproduce the published form of this manuscript, or allow others to do so, for United States Government purposes. The Department of Energy will provide public access to these results of federally sponsored research in accordance with the DOE Public Access Plan (www.energy.gov/doe-public-access-plan).

ACKNOWLEDGMENTS

The writing of this manuscript was supported by the Center for Bioenergy Innovation, a U.S. Department of Energy (DOE) Bioenergy Research Center supported by the Biological and Environmental Research (BER) program. Oak Ridge National Laboratory is managed by UT-Battelle, LLC for the U.S. Department of Energy under Contract Number DE-AC05-00OR22725. This article has been made freely available online by generous financial support from the Bill and Melinda Gates Foundation. We particularly want to acknowledge the support and encouragement of Rodger Voorhies, president, Global Growth and Opportunity Fund.

REFERENCES

Abraham PE, Yin H, Borland AM, Weighill D, Lim SD, De Paoli HC, Engle N, Jones PC, Agh R, Weston DJ, et al. 2016. Transcript, protein and metabolite temporal dynamics in the CAM plant agave. *Nat Plants* **2:** 16178. doi:10.1038/nplants.2016.178

Amin AB, Rathnayake KN, Yim WC, Garcia TM, Wone B, Cushman JC, Wone BWM. 2019. Crassulacean acid metabolism abiotic stress-responsive transcription factors: a potential genetic engineering approach for improving crop tolerance to abiotic stress. *Front Plant Sci* **10:** 129. doi:10.3389/fpls.2019.00129

Anand A, Bass SH, Wu E, Wang N, McBride KE, Annaluru N, Miller M, Hua M, Jones TJ. 2018. An improved ternary vector system for *Agrobacterium*-mediated rapid maize transformation. *Plant Mol Biol* **97:** 187–200. doi:10.1007/s11103-018-0732-y

Aregawi K, Shen J, Pierroz G, Sharma MK, Dahlberg J, Owiti J, Lemaux PG. 2022. Morphogene-assisted transformation of *Sorghum bicolor* allows more efficient genome editing. *Plant Biotechnol J* **20:** 748–760. doi:10.1111/pbi.13754

Bawa G, Liu Z, Yu X, Qin A, Sun X. 2022. Single-cell RNA sequencing for plant research: insights and possible benefits. *Int J Mol Sci* **23:** 4497. doi:10.3390/ijms23094497

Beilsmith K, Henry CS, Seaver SMD. 2022. Genome-scale modeling of the primary-specialized metabolism interface. *Curr Opin Plant Biol* **68:** 102244. doi:10.1016/j.pbi.2022.102244

Bertea CM, Scannerini S, Camusso W, Bossi S, Buffa G, Maffei M, D'Agostino G, Mucciarelli M. 2001. Evidence for a C$_4$ NADP-ME photosynthetic pathway in *Vetiveria zizanioides* Stapf. *Plant Biosyst* **135:** 249–262. doi:10.1080/11263500112331350890

Bian X, Jiang H, Meng Y, Li YP, Fang J, Lu Z. 2022. Regulation of gene expression by glycolytic and gluconeogenic enzymes. *Trends Cell Biol* **32:** 786–799. doi:10.1016/j.tcb.2022.02.003

Black CC, Osmond CB. 2003. Crassulacean acid metabolism photosynthesis: "working the night shift." *Photosynth Res* **76:** 329–341. doi:10.1023/A:1024978220193

Bläsing OE, Westhoff P, Svensson P. 2000. Evolution of C$_4$ phosphoenolpyruvate carboxylase in Flaveria, a conserved serine residue in the carboxyl-terminal part of the enzyme is a major determinant for C$_4$-specific char-

acteristics. *J Biol Chem* **275:** 27917–27923. doi:10.1074/jbc.M909832199

Borland AM, Yang X. 2013. Informing the improvement and biodesign of crassulacean acid metabolism via system dynamics modelling. *New Phytol* **200:** 946–949. doi:10.1111/nph.12529

Borland AM, Griffiths H, Hartwell J, Smith JAC. 2009. Exploiting the potential of plants with crassulacean acid metabolism for bioenergy production on marginal lands. *J Exp Bot* **60:** 2879–2896. doi:10.1093/jxb/erp118

Borland AM, Hartwell J, Weston DJ, Schlauch KA, Tschaplinski TJ, Tuskan GA, Yang X, Cushman JC. 2014. Engineering crassulacean acid metabolism to improve water-use efficiency. *Trends Plant Sci* **19:** 327–338. doi:10.1016/j.tplants.2014.01.006

Clark TJ, Guo L, Morgan J, Schwender J. 2020. Modeling plant metabolism: from network reconstruction to mechanistic models. *Annu Rev Plant Biol* **71:** 303–326. doi:10.1146/annurev-arplant-050718-100221

Cominelli E, Galbiati M, Tonelli C. 2010. Transcription factors controlling stomatal movements and drought tolerance. *Transcription* **1:** 41–45. doi:10.4161/trns.1.1.12064

Cushman JC, Tillett RL, Wood JA, Branco JM, Schlauch KA. 2008. Large-scale mRNA expression profiling in the common ice plant, *Mesembryanthemum crystallinum*, performing C$_3$ photosynthesis and crassulacean acid metabolism (CAM). *J Exp Bot* **59:** 1875–1894. doi:10.1093/jxb/ern008

Davis SC, LeBauer DS, Long SP. 2014. Light to liquid fuel: theoretical and realized energy conversion efficiency of plants using crassulacean acid metabolism (CAM) in arid conditions. *J Exp Bot* **65:** 3471–3478. doi:10.1093/jxb/eru163

Edwards EJ. 2019. Evolutionary trajectories, accessibility and other metaphors: the case of C$_4$ and CAM photosynthesis. *New Phytol* **223:** 1742–1755. doi:10.1111/nph.15851

Ferrari RC, Kawabata AB, Ferreira SS, Hartwell J, Freschi L. 2022. A matter of time: regulatory events behind the synchronization of C$_4$ and crassulacean acid metabolism in *Portulaca oleracea*. *J Exp Bot* **73:** 4867–4885. doi:10.1093/jxb/erac163

Friso G, Majeran W, Huang M, Sun Q, van Wijk KJ. 2010. Reconstruction of metabolic pathways, protein expression, and homeostasis machineries across maize bundle sheath and mesophyll chloroplasts: large-scale quantitative proteomics using the first maize genome assembly. *Plant Physiol* **152:** 1219–1250. doi:10.1104/pp.109.152694

Fujiki Y, Yoshikawa Y, Sato T, Inada N, Ito M, Nishida I, Watanabe A. 2001. Dark-inducible genes from *Arabidopsis thaliana* are associated with leaf senescence and repressed by sugars. *Physiol Plant* **111:** 345–352. doi:10.1034/j.1399-3054.2001.1110312.x

Furbank RT, Kelly S. 2021. Finding the C$_4$ sweet spot: cellular compartmentation of carbohydrate metabolism in C$_4$ photosynthesis. *J Exp Bot* **72:** 6018–6026. doi:10.1093/jxb/erab290

Gilman IS, Moreno-Villena JJ, Lewis ZR, Goolsby EW, Edwards EJ. 2022. Gene co-expression reveals the modularity and integration of C$_4$ and CAM in *Portulaca*. *Plant Physiol* **189:** 735–753. doi:10.1093/plphys/kiac116

Hafke JB, Hafke Y, Smith JAC, Lüttge U, Thiel G. 2003. Vacuolar malate uptake is mediated by an anion-selective inward rectifier. *Plant J* **35**: 116–128. doi:10.1046/j.1365-313X.2003.01781.x

Hassan MM, Zhang Y, Yuan G, De K, Chen JG, Muchero W, Tuskan GA, Qi Y, Yang X. 2021. Construct design for CRISPR/Cas-based genome editing in plants. *Trends Plant Sci* **26**: 1133–1152. doi:10.1016/j.tplants.2021.06.015

Hassan MM, Yuan G, Liu Y, Alam M, Eckert CA, Tuskan GA, Golz JF, Yang X. 2022. Precision genome editing in plants using gene targeting and prime editing: existing and emerging strategies. *Biotechnol J* **17**: 2100673. doi:10.1002/biot.202100673

Heyduk K, Moreno-Villena JJ, Gilman IS, Christin PA, Edwards EJ. 2019. The genetics of convergent evolution: insights from plant photosynthesis. *Nat Rev Genet* **20**: 485–493. doi:10.1038/s41576-019-0107-5

Huang S, Ding M, Roelfsema MRG, Dreyer I, Scherzer S, Al-Rasheid KAS, Gao S, Nagel G, Hedrich R, Konrad KR. 2021a. Optogenetic control of the guard cell membrane potential and stomatal movement by the light-gated anion channel GtACR1. *Sci Adv* **7**: eabg4619. doi:10.1126/sciadv.abg4619

Huang XY, Wang CK, Zhao YW, Sun CH, Hu DG. 2021b. Mechanisms and regulation of organic acid accumulation in plant vacuoles. *Hortic Res* **8**: 227. doi:10.1038/s41438-021-00702-z

Jain N, Khurana P, Khurana JP. 2022. A rapid and efficient protocol for genotype-independent, *Agrobacterium*-mediated transformation of *indica* and *japonica* rice using mature seed-derived embryogenic calli. *Plant Cell Tiss Organ Cult* **151**: 59–73. doi:10.1007/s11240-022-02331-3

Jia Y, Liu Y, Zhang S. 2021. Evaluation of agricultural ecosystem service value in arid and semiarid regions of northwest China based on the equivalent factor method. *Environ Process* **8**: 713–727. doi:10.1007/s40710-021-00514-2

Jones JJ, Huang S, Hedrich R, Geilfus CM, Roelfsema MRG. 2022. The green light gap: a window of opportunity for optogenetic control of stomatal movement. *New Phytol* **236**: 1237–1244. doi:10.1111/nph.18451

Kashtoh H, Baek KH. 2021. Structural and functional insights into the role of guard cell ion channels in abiotic stress-induced stomatal closure. *Plants* **10**: 2774. doi:10.3390/plants10122774

Khodayari A, Maranas CD. 2016. A genome-scale *Escherichia coli* kinetic metabolic model k-ecoli457 satisfying flux data for multiple mutant strains. *Nat Commun* **7**: 13806. doi:10.1038/ncomms13806

Kollist H, Nuhkat M, Roelfsema MRG. 2014. Closing gaps: linking elements that control stomatal movement. *New Phytol* **203**: 44–62. doi:10.1111/nph.12832

Kovermann P, Meyer S, Hörtensteiner S, Picco C, Scholz-Starke J, Ravera S, Lee Y, Martinoia E. 2007. The *Arabidopsis* vacuolar malate channel is a member of the ALMT family. *Plant J* **52**: 1169–1180. doi:10.1111/j.1365-313X.2007.03367.x

Li XD, Gao YQ, Wu WH, Chen LM, Wang Y. 2022. Two calcium-dependent protein kinases enhance maize drought tolerance by activating anion channel ZmSLAC1

in guard cells. *Plant Biotechnol J* **20**: 143–157. doi:10.1111/pbi.13701

Lim SD, Yim WC, Liu D, Hu R, Yang X, Cushman JC. 2018. A *Vitis vinifera* basic helix–loop–helix transcription factor enhances plant cell size, vegetative biomass and reproductive yield. *Plant Biotechnol J* **16**: 1595–1615. doi:10.1111/pbi.12898

Lim SD, Mayer JA, Yim WC, Cushman JC. 2020. Plant tissue succulence engineering improves water-use efficiency, water-deficit stress attenuation and salinity tolerance in *Arabidopsis*. *Plant J* **103**: 1049–1072. doi:10.1111/tpj.14783

Liu D, Hu R, Zhang J, Guo HB, Cheng H, Li L, Borland AM, Qin H, Chen JG, Muchero W, et al. 2021. Overexpression of an Agave phosphoenolpyruvate carboxylase improves plant growth and stress tolerance. *Cells* **10**: 582. doi:10.3390/cells10030582

Liu Z, Yu X, Qin A, Zhao Z, Liu Y, Sun S, Liu H, Guo C, Wu R, Yang J, et al. 2022. Research strategies for single-cell transcriptome analysis in plant leaves. *Plant J* **112**: 27–37. doi:10.1111/tpj.15927

Long CP, Antoniewicz MR. 2019. High-resolution ^{13}C metabolic flux analysis. *Nat Protoc* **14**: 2856–2877. doi:10.1038/s41596-019-0204-0

McGinn M, Phippen WB, Chopra R, Bansal S, Jarvis BA, Phippen ME, Dorn KM, Esfahanian M, Nazarenus TJ, Cahoon EB, et al. 2019. Molecular tools enabling pennycress (*Thlaspi arvense*) as a model plant and oilseed cash cover crop. *Plant Biotechnol J* **17**: 776–788. doi:10.1111/pbi.13014

Ming R, VanBuren R, Wai CM, Tang H, Schatz MC, Bowers JE, Lyons E, Wang ML, Chen J, Biggers E, et al. 2015. The pineapple genome and the evolution of CAM photosynthesis. *Nat Genet* **47**: 1435–1442. doi:10.1038/ng.3435

Moreno-Villena JJ, Zhou H, Gilman IS, Tausta SL, Cheung CYM, Edwards EJ. 2022. Spatial resolution of an integrated C_4+CAM photosynthetic metabolism. *Sci Adv* **8**: eabn2349. doi:10.1126/sciadv.abn2349

Nelson EA, Sage RF. 2008. Functional constraints of CAM leaf anatomy: tight cell packing is associated with increased CAM function across a gradient of CAM expression. *J Exp Bot* **59**: 1841–1850. doi:10.1093/jxb/erm346

Nelson-Vasilchik K, Hague JP, Tilelli M, Kausch AP. 2022. Rapid transformation and plant regeneration of sorghum (*Sorghum bicolor* L.) mediated by altruistic Baby boom and Wuschel2. *In Vitro Cell Dev Biol-Plant* **58**: 331–342. doi:10.1007/s11627-021-10243-8

Noronha H, Silva A, Dai Z, Gallusci P, Rombolà AD, Delrot S, Gerós H. 2018. A molecular perspective on starch metabolism in woody tissues. *Planta* **248**: 559–568. doi:10.1007/s00425-018-2954-2

Ochoa-Fernandez R, Abel NB, Wieland FG, Schlegel J, Koch LA, Miller JB, Engesser R, Giuriani G, Brandl SM, Timmer J, et al. 2020. Optogenetic control of gene expression in plants in the presence of ambient white light. *Nat Methods* **17**: 717–725. doi:10.1038/s41592-020-0868-y

Okubo-Kurihara E, Sano T, Higaki T, Kutsuna N, Hasezawa S. 2009. Acceleration of vacuolar regeneration and cell growth by overexpression of an aquaporin NtTIP1;1 in tobacco BY-2 cells. *Plant Cell Physiol* **50**: 151–160. doi:10.1093/pcp/pcn181

Omelina ES, Yushkova AA, Motorina DM, Volegov GA, Kozhevnikova EN, Pindyurin AV. 2022. Optogenetic and chemical induction systems for regulation of transgene expression in plants: use in basic and applied research. *Int J Mol Sci* **23:** 1737. doi:10.3390/ijms23031737

Owen NA, Griffiths H. 2013. A system dynamics model integrating physiology and biochemical regulation predicts extent of crassulacean acid metabolism (CAM) phases. *New Phytol* **200:** 1116–1131. doi:10.1111/nph.12461

Pan C, Sretenovic S, Qi Y. 2021a. CRISPR/dCas-mediated transcriptional and epigenetic regulation in plants. *Curr Opin Plant Biol* **60:** 101980. doi:10.1016/j.pbi.2020.101980

Pan C, Wu X, Markel K, Malzahn AA, Kundagrami N, Sretenovic S, Zhang Y, Cheng Y, Shih PM, Qi Y. 2021b. CRISPR-act3.0 for highly efficient multiplexed gene activation in plants. *Nat Plants* **7:** 942–953. doi:10.1038/s41477-021-00953-7

Pandey S, Zhang W, Assmann SM. 2007. Roles of ion channels and transporters in guard cell signal transduction. *FEBS Lett* **581:** 2325–2336. doi:10.1016/j.febslet.2007.04.008

Peterson D, Barone P, Lenderts B, Schwartz C, Feigenbutz L, St. Clair G, Jones S, Svitashev S. 2021. Advances in *Agrobacterium* transformation and vector design result in high-frequency targeted gene insertion in maize. *Plant Biotechnol J* **19:** 2000–2010. doi:10.1111/pbi.13613

Prías-Blanco M, Chappell TM, Freed EF, Illa-Berenguer E, Eckert CA, Parrott WA. 2022. An *Agrobacterium* strain auxotrophic for methionine is useful for switchgrass transformation. *Transgenic Res* **31:** 661–676. doi:10.1007/s11248-022-00328-4

Reyna-Llorens I, Aubry S. 2022. As right as rain: deciphering drought-related metabolic flexibility in the C_4–CAM *Portulaca. J Exp Bot* **73:** 4615–4619. doi:10.1093/jxb/erac179

Rojas BE, Hartman MD, Figueroa CM, Leaden L, Podestá FE, Iglesias AA. 2019. Biochemical characterization of phospho*enol*pyruvate carboxykinases from *Arabidopsis thaliana. Biochem J* **476:** 2939–2952. doi:10.1042/BCJ20190523

Sage RF. 2002. Are crassulacean acid metabolism and C_4 photosynthesis incompatible? *Funct Plant Biol* **29:** 775–785. doi:10.1071/PP01217

Santelia D, Lawson T. 2016. Rethinking guard cell metabolism. *Plant Physiol* **172:** 1371–1392. doi:10.1104/pp.16.00767

Schiller K, Bräutigam A. 2021. Engineering of crassulacean acid metabolism. *Annu Rev Plant Biol* **72:** 77–103. doi:10.1146/annurev-arplant-071720-104814

Shameer S, Baghalian K, Cheung CYM, Ratcliffe RG, Sweetlove LJ. 2018. Computational analysis of the productivity potential of CAM. *Nat Plants* **4:** 165–171. doi:10.1038/s41477-018-0112-2

Simons M, Misra A, Sriram G. 2014. Genome-scale models of plant metabolism. In *Plant metabolism: methods and protocols* (ed. Sriram G), pp. 213–230. Humana, Totowa, NJ.

Song C, Lu L, Guo Y, Xu H, Li R. 2019. Efficient *Agrobacterium*-mediated transformation of the commercial hybrid poplar *Populus alba* × *Populus glandulosa* Uyeki. *Int J Mol Sci* **20:** 2594. doi:10.3390/ijms20102594

Sussmilch FC, Roelfsema MRG, Hedrich R. 2019. On the origins of osmotically driven stomatal movements. *New Phytol* **222:** 84–90. doi:10.1111/nph.15593

Tay IYY, Odang KB, Cheung CYM. 2021. Metabolic modeling of the C_3-CAM continuum revealed the establishment of a starch/sugar-malate cycle in CAM evolution. *Front Plant Sci* **11:** 573197. doi:10.3389/fpls.2020.573197

Tian C, Zhang Y, Li J, Wang Y. 2022. Benchmarking intrinsic promoters and terminators for plant synthetic biology research. *BioDesign Res* **2022:** 9834989. doi:10.34133/2022/9834989

Wang PH, Kumar S, Zeng J, McEwan R, Wright TR, Gupta M. 2020. Transcription terminator-mediated enhancement in transgene expression in maize: preponderance of the AUGAAU motif overlapping with poly(A) signals. *Front Plant Sci* **11:** 570778. doi:10.3389/fpls.2020.570778

Wen SS, Ge XL, Wang R, Yang HF, Bai YE, Guo YH, Zhang J, Lu MZ, Zhao ST, Wang LQ. 2022. An efficient *Agrobacterium*-mediated transformation method for hybrid poplar 84K (*Populus alba* × *P. glandulosa*) using calli as explants. *Int J Mol Sci* **23:** 2216. doi:10.3390/ijms23042216

Winter K. 2019. Ecophysiology of constitutive and facultative CAM photosynthesis. *J Exp Bot* **70:** 6495–6508. doi:10.1093/jxb/erz002

Winter K, Holtum JAM. 2014. Facultative crassulacean acid metabolism (CAM) plants: powerful tools for unravelling the functional elements of CAM photosynthesis. *J Exp Bot* **65:** 3425–3441. doi:10.1093/jxb/eru063

Winter K, Smith JAC. 2022. CAM photosynthesis: the acid test. *New Phytol* **233:** 599–609. doi:10.1111/nph.17790

Xia K, Sun HX, Li J, Li J, Zhao Y, Chen L, Qin C, Chen R, Chen Z, Liu G, et al. 2022. The single-cell stereo-seq reveals region-specific cell subtypes and transcriptome profiling in *Arabidopsis* leaves. *Dev Cell* **57:** 1299–1310.e4. doi:10.1016/j.devcel.2022.04.011

Xu N, Kang M, Zobrist JD, Wang K, Fei SZ. 2022. Genetic transformation of recalcitrant upland switchgrass using morphogenic genes. *Front Plant Sci* **12:** 781565. doi:10.3389/fpls.2021.781565

Yang X, Cushman JC, Borland AM, Edwards EJ, Wullschleger SD, Tuskan GA, Owen NA, Griffiths H, Smith JAC, De Paoli HC, et al. 2015. A roadmap for research on crassulacean acid metabolism (CAM) to enhance sustainable food and bioenergy production in a hotter, drier world. *New Phytol* **207:** 491–504. doi:10.1111/nph.13393

Yang X, Hu R, Yin H, Jenkins J, Shu S, Tang H, Liu D, Weighill DA, Cheol Yim W, Ha J, et al. 2017. The Kalanchoë genome provides insights into convergent evolution and building blocks of crassulacean acid metabolism. *Nat Commun* **8:** 1899. doi:10.1038/s41467-017-01491-7

Yang X, Liu D, Tschaplinski TJ, Tuskan GA. 2019. Comparative genomics can provide new insights into the evolutionary mechanisms and gene function in CAM plants. *J Exp Bot* **70:** 6539–6547. doi:10.1093/jxb/erz408

Yang X, Medford JI, Markel K, Shih PM, De Paoli HC, Trinh CT, McCormick AJ, Ployet R, Hussey SG, Myburg AA, et al. 2020. Plant biosystems design research roadmap 1.0. *BioDesign Res* **2020:** 8051764. doi:10.34133/2020/8051764

Yang X, Liu D, Lu H, Weston DJ, Chen JG, Muchero W, Martin S, Liu Y, Hassan MM, Yuan G, et al. 2021. Biological parts for plant biodesign to enhance land-based carbon dioxide removal. *BioDesign Res* **2021**: 9798714. doi:10.34133/2021/9798714

Yin H, Guo HB, Weston DJ, Borland AM, Ranjan P, Abraham PE, Jawdy SS, Wachira J, Tuskan GA, Tschaplinski TJ, et al. 2018. Diel rewiring and positive selection of ancient plant proteins enabled evolution of CAM photosynthesis in *Agave*. *BMC Genomics* **19**: 588. doi:10.1186/s12864-018-4964-7

Yuan G, Hassan MM, Liu D, Lim SD, Yim WC, Cushman JC, Markel K, Shih PM, Lu H, Weston DJ, et al. 2020. Biosystems design to accelerate C$_3$-to-CAM progression. *BioDesign Res* **2020**: 3686791. doi:10.34133/2020/3686791

Yuan G, Lu H, Tang D, Hassan MM, Li Y, Chen JG, Tuskan GA, Yang X. 2021. Expanding the application of a UV-visible reporter for transient gene expression and stable transformation in plants. *Hortic Res* **8**: 234. doi:10.1038/s41438-021-00663-3

Yuan G, Lu H, De K, Hassan MM, Wellington M, Tuskan GA, Yang X. 2022a. Split selectable marker mediated gene stacking in plants. bioRxiv doi:10.1101/2022.11.19.517202

Yuan G, Martin S, Hassan MM, Tuskan GA, Yang X. 2022b. PARA: a new platform for the rapid assembly of gRNA arrays for multiplexed CRISPR technologies. *Cells* **11**: 2467. doi:10.3390/cells11162467

Zhang J, Hu R, Sreedasyam A, Garcia TM, Lipzen A, Wang M, Yerramsetty P, Liu D, Ng V, Schmutz J, et al. 2020. Light-responsive expression atlas reveals the effects of light quality and intensity in *Kalanchoë fedtschenkoi*, a plant with crassulacean acid metabolism. *Gigascience* **9**: giaa018. doi:10.1093/gigascience/giaa018

Zhu R, Fu Y, Zhang L, Wei T, Jiang X, Wang M. 2022. A light-inducible bidirectional promoter initiates expression of both genes *SHH2* and *CFM3* in *Brassica napus* L. *J Plant Biol* doi:10.1007/s12374-022-09367-0

Transformation of Plantation Forestry Productivity for Climate Change Mitigation and Adaptation

Mike May, Stanley Hirsch, and Miron Abramson

FuturaGene Israel Ltd., Rehovot 76100, Israel

Correspondence: mike@futuragene.com

The protection of natural forests as the major land-based biotic sink of carbon is regarded as a priority for climate action, and zero deforestation is an accepted global imperative. Sustainable intensification of plantation forestry will be essential to meet escalating, shifting, and diversifying demand for forest products if logging pressure on natural forests is to be decreased. Substitution strategies involves enhanced offtake from plantation forestry into long life-cycle products, opening up new options for medium- to long-term carbon drawdown, downstream decarbonization, and fossil fuel displacement in the construction and chemicals sectors. However, under current plantation productivity levels, it has been projected that by 2050, supply could provide as little as 35% of demand. This could be further exacerbated by climate change. To mitigate this shortfall, to avoid ensuing catastrophic logging pressure on natural forests, and to ensure that downstream decarbonization and fossil fuel substitution strategies are feasible, a dramatic step change in plantation productivity is required. This is particularly necessary in developing countries where increases in per capita demand and pressure on natural forests will be the most acute.

Over the last 20 years, gene modification, gene editing, and genomics approaches for yield enhancement and yield protection have been developed for eucalyptus, one of the most significant plantation species globally. This work delivers positive insights for carbon drawdown and for driving climate change mitigation, adaptation, and resilience strategies. We present here a snapshot of our experience to date in Brazil of genetically modified eucalyptus enhanced for greater yield (carbon drawdown), and photorespiratory bypass (climate adaptation). In parallel, we have developed a robust governance framework based on a number of ethical principles and practical precautionary steps, including global dialogue and technology sharing. Here we describe possible ways forward to explore carbon drawdown at scale and climate resilience through field experimentation and discuss the potential future impact of these approaches. One of our primary objectives is to use landscape-level experimentation as a platform for global dialogue to create a new narrative around biotech plantations on the governance required to unlock technologies for the benefit of climate change mitigation, biodiversity, and equality.

Cite this article as *Cold Spring Harb Perspect Biol* doi: 10.1101/cshperspect.a041670

While many of the technologies under development hold considerable greenhouse gas (GHG) mitigation potential, the rate-limiting factor in their ability to sequester carbon is public acceptance.

THE PRODUCTIVITY GAP

As demand for food, fiber, and fuel increases, wood harvesting will triple by 2050 to ~10 billion m^3 per annum, which would require a doubling of the existing plantation area: roughly 250 million ha of new plantations under existing productivity levels (World Wildlife Federation 2011). More recent estimates indicate that timber consumption for construction alone is predicted to triple from 2.2 billion m^3 to 5.8 billion m^3 in 2050, driven by urbanization and political commitments to decarbonization (Gresham House, greshamhouse.com/global-timber-outlook). Furthermore, over the last 20 years, the source of demand has changed radically. The developing world has overtaken the developed world, growing from 770 million m^3 in 1998 to 1.2 billion m^3 in 2018, with China leading this shift as a result of population growth and urbanization. Wood consumption is tightly linked to population growth; with a further 800 million people expected by 2050—two-thirds of which will be in Africa—this trend will accelerate. However, today, the vast majority of timber (74%) is produced in the global North (United States, Canada, Europe, Russia). The only solution to avert massive deforestation in the global South is therefore significant upscaling of local production capacity. For some countries, this means starting organized forestry operations from an almost zero base. This was recently underlined by Sir David Attenborough, describing the use of more wood as "an extraordinary renewable resource" for growing populations—"On their own, natural forests cannot provide all the wood we need so we also have to farm trees like any other crop and create a new generation of plantations" (Attenborough 2020).

However, this forecasted rise in timber demand is set against a constrained supply fixed by the long-term growth rate of the trees, limited land area, climate change, and an uncertain investment scenario. Current forecasts for long-term timber supply between now and 2050 range between 3.7 billion m^3 and 4.7 billion m^3, representing a considerable shortfall. The International Tropical Timber Organisation (ITTO, www.itto.int) estimates even greater shortfall caused by inadequate plantation expansion globally. The total area of industrial fast-growing forest plantations in 2012 was 54.3 million ha and this area will increase to 91 million ha by 2050, satisfying only 35% of total industrial wood requirements (ITTO 2020).

Addressing this projected deficit places considerable pressure on the global forest sector to dramatically increase productivity while reducing logging pressure on natural forests and also scaling up landscape restoration activities (IUCN 2009). Global momentum to decarbonize economies is expected to further accelerate timber consumption, as wood is increasingly seen as a clean and low carbon alternative to fossil fuel and carbon intensive inputs (European Commission 2020). Furthermore, increased fire risk, increased amplitude, and duration of pest and disease cycles, water constraints, and impaired photosynthesis as a result of climate change will require the development and deployment of new and effective mitigation strategies that increase productivity at scale and on time.

FIXING THE SUPPLY–DEMAND GAP BY 2050—INCREASING THE PRODUCTIVITY OF PLANTATIONS

Plantations have for many years been seen as part of the solution (Indufor 2012) for reducing logging pressure on natural forests, while meeting growing, diversifying, and shifting demand. Under appropriate landscape-level zoning, local community consultation, and the appropriate genetics, plantations can produce more wood on less land than natural forest, and could thus spare land for other uses (newgenerationplantations.org). Estimates underline that while tree plantations comprise only 7% of total forest area, they provide 50% of industrial roundwood (Jagels 2006). Of these, the most productive plantations—intensively managed planted forests (IMPFs)—cover only 0.2% of global land area

Cite this article as *Cold Spring Harb Perspect Biol* doi: 10.1101/cshperspect.a041670

or 1.25% of global forest cover, but provide 30% of global roundwood production (Kanowski and Murray 2008). Further intensification is, however, needed to mitigate projected supply deficits, and is entirely feasible since in many regions, plantations are not managed at their optimal productivity. Tripling the yield and harvest of planted forests from 800 million m^3 from 7% of the world's forests to 2.7 billion m^3 from 11% of the world's forests would create savings of 2.35 Gt CO_2/yr and a net gain of 1.5 Gt CO_2/yr (Flores Carreño et al. 2012). Improving and protecting yields of planted forests would be achieved "through genetic improvements that emphasize a mix of plant traits (drought tolerance, insect resistance, product characteristics) and adaptation to different forest types and locations" (Flores Carreño et al. 2012).

Fast-growing eucalyptus plantations have many of the characteristics required to meet part of this supply gap. While the eucalyptus commercial plantation area represents only ~0.5% of the world's forest cover (25 million ha globally [FAO 2020]), it accounts for ~10% of the global roundwood demand. Brazil is the world's largest eucalyptus grower (22%) (followed by China [20%] and India [17%]) (Wen et al. 2018), with 7.6 million ha, constituting ~7% of Brazil's GDP (IBÁ 2022).

Brazil, more importantly, is by far the leader in productivity, with an average mass accumulation of 40 $m^3ha^{-1}yr^{-1}$, resulting from a decades-long history of investment in breeding and improvements in silvicultural practices. This has created a diverse clonal base of varieties that are adaptable to many biomes and relatively tolerant to biotic and abiotic stress (da Silva et al. 2013; Ramantswana et al. 2020). Long-term studies have demonstrated that the eucalyptus plantations are sustainable, and very efficient at capturing CO_2 (240 tons of CO_2 per ha over a 7-year rotation (IBÁ 2022). Furthermore, under the norms of the Brazilian Forest Code, landowners may only establish plantations on degraded land and must commit between 20% and 50% of their land to native forest renewal, making this industry one of the global leaders in degraded land restoration (Grossberg 2009; Sembiring et al. 2020; Coelho et al. 2022).

CREATING STEP CHANGES IN BRAZILIAN PLANTATION FORESTRY

Since the 1970s, eucalyptus productivity in Brazil has more than doubled from 17 m^3 to 40 m^3ha^{-1} yr^{-1} through access to improved germplasm, better breeding techniques, clonal propagation, improved silviculture, and improved plantation management practices. Breeding for better performing tree varieties provides continuous improvement in quality and supply while maintaining genetic diversity and continues to drive down the amount of land needed to produce a given amount of pulp: to produce 500,000 tons of pulp per year in the 1980s, a pulp mill needed ~67 k ha of land, ~51% more than that needed today (~45 k ha), to supply the same amount of pulp annually.

However, yield improvements through conventional approaches are nearing a ceiling. In addition, some traits cannot be created through conventional breeding (such as insect resistance or modified wood properties), and for some traits, such as climate resilience, conventional breeding time lines are too long. If enhancing plantation productivity requires a continued emphasis on raising technical efficiency, then genetic modification (GM) could be part of this solution. Since 2006, FuturaGene has been integrating GM within conventional breeding programs, affording the potential to enhance and protect yield, to modify wood properties for diverse offtakes, and to prepare for climate change while using less land and resources and reduce chemical loads. This approach holds great promise for enabling the essential step change in productivity and intensification of existing practices for closing the supply gap. As of March 2023, FuturaGene has received seven approvals for the commercial deployment of biotech traits in eucalyptus from the Brazilian Technical Biosafety Commission (CTNBio). The approvals include a yield enhancement trait, five for herbicide resistance and an insect resistance trait. All these traits are aimed at enhancing and protecting plantation productivity with lower inputs and reduced fossil fuel demand for operations.

Underpinning our scientific and regulatory capabilities, FuturaGene and our parent company Suzano have developed a robust governance system to enable the implementation of relevant tree biotechnology approaches as part of the solution to mitigating global existential crises of climate change, food security, biome loss, and equality. A central part of this governance system is a detailed corporate GM tree policy accessible here: storage.googleapis.com/stateless-site-suzano-en/ 2021/06/a7bf592a-suzano-gm-tree-policy-_-final_ 17.06.21-reviewed.pdf. This policy is based on principles and commitments that govern Suzano's GM tree program and relationships with stakeholders. These include:

1. A scientific and operational framework to ensure all projects provide solutions that enable not only the transformation of the economic values of forestry but also create social shared values and environmental renewal and resilience; and

2. A bioethical framework to ensure decisions shall be based on dialogue with stakeholders, on the principle of shared values, and on an understanding of direct and indirect environmental impacts and dependencies. This framework would be implemented through commitments to ensure transparency, stewardship, and traceable good practices within all operations; stimulate long-term, outcome-oriented, local and global, inclusive dialogues on tree biotechnology; promote technology transfer for humanitarian or environmental purposes; align all initiatives with opportunities to share benefits with outgrowers (contracted small-holder farmers); and implement a bioethics advisory committee to identify and define technical, scientific, and bioethical questions about the social, environmental, economic, and legal aspects of biotechnology and its applications.

These commitments create the potential for community empowerment, resilience, and entrepreneurism at a local level, and globally through sharing our technologies free of charge to public sector research institutes working on the improvement of subsistence crops for food security or institutes developing strategies for safeguarding forest health. To date, we have shared our technologies with the Donald Danforth Center to improve subsistence crops (Ven-

kata et al. 2020); we have provided a eucalyptus research clone and genetic transformation protocols to a research consortium based at Oregon State University; and we have put the most advanced eucalyptus genome data into the public domain to advance public research (FuturaGene, www.futuragene.com/shared-value). Furthermore, during in-licensing of key technologies, we ensure that smallholder outgrowers of Suzano have access to technologies free of any royalty charges.

We describe here three approaches to scale-up and expand our programs to reframe global thinking on carbon drawdown, closing the supply gap, climate resilience, and to reframe critical questions regarding the relevance of tree biotechnology for society and the environment.

LANDSCAPE-LEVEL EXPERIMENTATION AND GLOBAL DIALOGUE TO CHANGE THE NARRATIVE ON TREE BIOTECHNOLOGY

Assessing Carbon Assimilation in Yield-Enhanced, GM Eucalyptus Plantations at a Landscape Level in Brazil

FuturaGene has developed a genetically modified yield-enhanced eucalyptus event generated by the transgenic expression of a cell wall-modifying gene (1,4-β-endoglucanase) from *Arabidopsis* that allows enhanced cell elongation (Shani et al. 2004; Abramson et al. 2013). The event was approved for commercial use by the CTNBio in April 2015 (Nature Biotechnology 2015).

Our field experiments to date have yielded a large body of evidence on biosafety and performance criteria that demonstrates that our transgenic approach delivers substantial yield enhancement in different genetic (G) backgrounds in different environments (E). In fact, extended field trials with over 500 different clones, involving over 300,000 trees in different growing areas around Brazil since 2006, have shown conclusively that this technology affords significantly higher productivity of wood per planted area over the best available commercial clones in specific biomes (G×E effect). Although linear incremental improvements in yield of Brazilian eucalyptus have been achieved since the onset of breeding programs in the 1970s, increments achieved through conventional breeding are stagnating. The gains from the genetically modified yield enhanced eucalyptus event in specific biomes can represent up to almost half of the entire yield gain achieved over the last 40 years of conventional breeding. Thus, our yield-enhancing technology offers the potential to change the paradigm of yield enhancement and directly address the yield gap.

Currently, more than 6 million ha are planted with eucalyptus in Brazil alone. New plantings are only permitted on degraded pastureland, which is a very poor carbon sink. Data from the IBÁ (the Brazilian Tree Association) indicates that conventional eucalyptus may sequester 240 tons of CO_2 per ha over a 7-year rotation (IBÁ 2022). An average increase of 12% in yield would potentially push this to 270 tons of CO_2 per ha. Furthermore, it is estimated that ~30% of the biomass generated during a 7-year growth cycle remains in the land, through thinnings, root mass, which is not removed, and debarking in the field, which is also a major source of remineralization of the land and soil carbon regeneration.

What is needed now to better understand the carbon drawdown potential of these trees is landscape-level experimentation in different biomes, including microbial assimilation and soil organic carbon dynamics. Here, we propose to evaluate carbon assimilation of the yield-enhanced event over the next 6 years (a typical rotation) at landscape scale (1000 to 3000 ha) to create a quantitative baseline for decision making and gain a global understanding of the impact of landscape-level planting of modified eucalyptus on atmospheric carbon drawdown prior to harvest.

Landscape-level analyses will deliver solid baseline data for assessing carbon drawdown in different G×E combinations for the first time and would provide detailed landscape-level data on the potential to enhance the carbon drawdown capability of this crop. Data derived from such experimentation could be of major significance for decision making with respect to carbon drawdown implications in Brazil in particular, with a planted eucalyptus base of 6 million ha, as well as with respect to plantation forestry globally. This approach could also serve as a blueprint for similar studies in other parts of the eucalyptus-growing regions in the tropics and subtropics.

Metabolic Engineering of Photorespiration Bypass in Transgenic Eucalyptus: Implications for Sustainability under Climate Change

Even with significant advances in the form of clonal adaptations, improved management practices, soil preparation and fertilization, and pest and disease control, eucalyptus productivity in Brazil is strongly affected by climate variability (Binkley et al. 2017; Martins et al. 2022). Water deficits (Scolforo et al. 2019; Martins et al. 2022) and extreme temperatures outside of appropriate thresholds (from 8.5°C to ~40°C) (de Freitas and Bolzan Martins 2019) are the main causes of reduced eucalyptus productivity in Brazil, and these impacts are expected to become more prevalent with climate change (Elli et al. 2020; Florencio et al. 2022). A series of approaches allow us to prepare for such an impact to provide varieties that will be able to perform well even under future adverse climate scenarios.

One approach that has already shown promise in other species (see Erb 2023) is to investigate ways to bypass photorespiration (Betti et al. 2016; South et al. 2018). During photosynthesis, the enzyme rubisco incorporates carbon dioxide into an organic molecule during carbon fixation, but rubisco has a flaw: instead of using carbon dioxide as a sole substrate, it sometimes consumes oxygen, particularly under conditions of high light and high temperature. This side reaction triggers a process called photorespiration, which causes carbon dioxide to be lost rather than fixed. Photorespiration (also known as the oxidative photosynthetic carbon cycle, photorespiratory cycle or C_2 cycle) is a complex and energy-consuming set of reactions required to recycle 2-phosphoglycolate generated by the oxygenase reaction of rubisco (Betti et al. 2016). Photorespiration wastes energy, hampers primary metabolism and carbon fixation, and produces toxic byproducts that can impair crop productivity by as much as 50% under severe conditions such as elevated temperatures. Increased temperatures enhance photorespiration, and with the current global warming, the negative impact of photorespiration on yield loss will increase. Recent studies indicate that plant yield can be substantially improved by the alteration of photorespiratory fluxes or by engineering artificial bypasses to photorespiration. To address these challenges and improve the sustainability of eucalyptus growth and cultivation, FuturaGene has developed a transgenic eucalyptus variety expressing a photorespiration bypass. The engineered photorespiration bypass would allow the transgenic eucalyptus to be more effective in capturing CO_2 and water, thus reducing the negative impact of photorespiration. Through monitoring gas-exchange parameters in the youngest fully expanded leaves, our findings show that the transgenic eucalyptus plants exhibit a significant increase in leaf photosynthetic capacity (measured as Amax under ambient conditions), ranging from 15% to 25% higher than nontransgenic plants. Moreover, we sought to investigate whether the enhanced photosynthetic performance of the transgenic plants resulted in any phenotypic differences. Our experiments with eucalyptus trees grown in greenhouse conditions revealed that the transgenic lines had an increased growth rate and demonstrated an increased height of up to 15% compared to nontransgenic trees, with no change in stem diameter. Although our experiments are limited to greenhouse studies at this stage, we have demonstrated that eucalyptus engineered with photorespiration bypass results in enhanced growth and increased carbon assimilation compared with nontransgenic controls. To gain a more comprehensive understanding of the effects of photorespiratory bypass on plant metabolism and performance, we plan further studies to investigate the impact of this GM on eucalyptus primary metabolites, specifically carbohydrates and amino acid composition. In addition, we plan to evaluate the performance of the transgenic eucalyptus trees under drought conditions.

This research emphasizes the potential for transgenic metabolic engineering to improve the sustainability and resilience of eucalyptus in the face of future climate change and to permit better strategies for forestry planning, silvicultural practices, and guiding effective adaptive measures to cope with climate change threats, especially in regions facing water scarcity and high temperatures. Further research is required to fully understand the potential implications of this technology and its feasibility for commercial use. The next step is to take these experiments to the field to assess the impact of the bypass under different genetic and environmental combinations in different biomes in Brazil.

Global Dialogue to Unlock the Potential of Biotechnology

Global development is on a trajectory of ecological deficit, inequality, and food insecurity, affecting the lives of billions and the prospects of future generations. Credible pathways to redirect global development are flawed by the disconnect between science and global development policy. Paradoxically, it is clear that the rate of scientific and technological innovation can keep pace with global developmental challenges—but only if it is allowed to do so.

A new narrative is needed that unlocks technology—global discourse needs to change from "how do you construct social processes to regulate biotechnology" to "how do you unlock the innovation required to take global development to an inclusive, low-carbon, and just trajectory?" Dialogue is essential to create a favorable operating environment that allows the implementation of science-based strategies to deliver the step change needed in the way land is used to achieve greater efficiencies of forest commodity production. It is clear that to achieve greater intensification of productivity, existing performance standards, designed to manage linear incremental change, will not suffice. Global dialogue must focus on the ways in which future standards and governance frameworks can be designed to manage the complexity of systemic transformational changes.

Leadership is needed in the formulation of these frameworks that will master production efficiency in transformative ways and manage the highly disruptive and controversial process of further intensification and that stimulates preferential procurement and increased consumer awareness.

For more than 10 years, FuturaGene has linked their scientific and regulatory program to a strong presence in global dialogue at all levels to reframe critical questions (May and Hirsch 2015). The approach has been to urge a discourse and collective reasoning on governance, and break the deadlock caused by decades of exchanges around risk. With a perspective of governance, it is possible to reframe what is essentially a "wicked problem" in terms of broader value for society and the environment. Our narrative has been framed on the premise that with limited scope for sustainable throughput of resources, ensuring well-being within planetary boundaries, will require enhanced resource use efficiency while meeting growing and shifting demand. In practical terms, this requires step changes in productivity and process efficiency and an expansion of the quality, scope, and scale of products and services, and that in turn means an increased dependence on scientific and technological innovation. The question we have posed is: "If the only way to achieve natural resource use efficiency and produce more from less is to improve technical efficiency—how can inclusive frameworks of governance for the development and deployment of scientific and technological innovation be established and implemented down to the local level?"

In practical terms: "How can research be guided toward productivity challenges, how can its impact reach smallholders, and how can they use it to access new markets?"

Our GM tree policy is designed specifically to operationalize this ethic through sharing technology with small producers in Brazil, sharing technology free of charge to public sector research institutes for humanitarian and environmental objectives (Venkata et al. 2020; FuturaGene 2023), and the establishment of a bioethics advisory committee to oversee the entire program.

We have made some progress in creating awareness around positive narratives, but further progress at scale and on time is essential. The large-scale field experiments we propose to assess carbon drawdown and yield enhancement provide a unique opportunity to expand and enrich this dialogue. Because of the scale of this approach and its novelty and relevance to the global debate on different approaches to carbon drawdown, we

plan to open the field experiments as a global platform for dialogue with the participation of the key institutions leading the global carbon drawdown and decarbonization discussions and key downstream industries through a series of workshops, trainings, and debates in situ. Such interactions would allow decision makers to see first-hand how carbon drawdown would work in practice. This would fill an important gap in the global "carbon drawdown" debate framed in the context of "what do real-life solutions actually look like?" The program would also involve field visits to protected forests, visits with local communities and social programs—and would build awareness that high-tech plantation forestry is a necessary option for development.

Workshops based on real-life experiences framed around critical discussion themes in situ would help bridge many current gaps in thought leadership and would provide a fact-based vehicle for introducing the role of tree biotechnology into global thinking. A critical objective will be to establish key facts on plantation management, to create awareness that well-placed and well-managed plantations can be an asset for local community well-being, biodiversity conservation, ecological health, and commercial benefit.

The Brazilian context in which our work evolves is unique in the world and offers a vision of how plantation forestry can provide a strong vehicle for rural development, biodiversity conservation, economic growth, and ecological regeneration (IBÁ 2022). By changing the content and direction of debate around the role of biotech forestry in carbon drawdown and the potential to narrow the supply gap at scale and on time, a new vision for plantation-based solutions for sustainability and developmental issues could emerge. Until there is a better understanding that plantations are part of the solution, not the problem, the investments needed to boost productivity in the subtropics (and enhance carbon drawdown and social values) will not be made.

We would use the same workshop-based approach to showcase the value of yield-enhanced, GM eucalyptus on narrowing the global supply–demand deficit and stimulate debate on the possible future impacts of the existing and future supply–demand crisis by direct experience in situ. Debate is urgently needed, particularly on the humanitarian aspect of this issue, and our project could provide a unique convening platform for deliberation of practical solutions.

DISCUSSION

There is growing consensus that a considerable shortfall between supply and demand for forest products will develop in coming decades (Indufor 2012; ITTO 2020). Planted trees already provide a significant proportion of supply, but at current planting and productivity levels, this cannot suffice to fill the gap. Planted trees also provide both a means for carbon drawdown (IBÁ 2022), avoiding deforestation (New Generation Plantations, new generationplantations.org), and a valuable feedstock for decarbonization and reducing emissions from downstream carbon and energy-intensive industries. If the global imperative of zero deforestation as a priority for climate action is to be achieved, the supply gap urgently needs technical solutions in the ground now and appropriate governance frameworks to guide their development and deployment. The need to intensify the productivity of plantations becomes a significant priority, because it is here that innovation will make a difference. The potential for positive cascading effects of decarbonizing the construction sector or substituting fossil fuels in the chemicals industry will not be possible if the supply of alternative sources—biomass—are not reliable or priced competitively against incumbent feedstocks.

Our research has shown that GM for yield enhancement and yield protection traits in eucalyptus can deliver step changes in productivity and climate resilience and are a first step toward radically improving plantation productivity. Nevertheless, there remains a significant lag between scientific advances and enabling policy that directs research toward sustainability objectives and ensures that ensuing products are disseminated widely and equitably. Access to and widespread use of such technologies can be enabled when appropriate policies that drive technology transfer, local capacity building, and preferential procurement are in place. Underpinning all of this will be greater public accep-

tance of the technologies involved and plantations themselves.

Such technologies and governance frameworks are urgently needed since the world is in ecological deficit and GDP growth remains tightly linked to environmental degradation, biodiversity loss, inequality, and carbon emissions (www.overshootday.org). Current patterns of land and resource use and the scale and intensity of land-use change are already radically altering the ecological infrastructure of the planet: 82% of 94 core ecological processes recognized by biologists, from genetic diversity to ecosystem function, are impaired (Scheffers et al. 2016).

Goals to deliver solutions for climate change, biodiversity loss, food security, and equality are imperatives since they are at or near tipping points. To date, however, the institutional arrangements established since the Earth Summit in 1992 have struggled to manage the complexity of these demands: overshoot of natural resource consumption, inequality, and degradation continue despite global commitments to create a low-carbon, inclusive, and just trajectory for global development. The UN 2030 Agenda of Sustainable Development Goals (SDGs), adopted in 2015, was already off-target across many indicators within years of its inception (World Economic Forum 2021); most indicators for preventing excessive warming are off-target (World Resources Institute 2021). In the interim, the escalating imbalance between human activity and natural resource capacities will continue. There is an obvious and disconcerting mismatch between the urgency with which the objectives of ensuring well-being within planetary boundaries and the time frames in which practical transformative forest solutions can be delivered. Radical, systemic action is required and applies to all sectors and countries at all scales since the problems addressed are tightly interlinked, evolving unpredictably and are systemic. Actions for climate change mitigation must be taken in parallel with actions to provide for biodiversity regeneration, food security, and equality. Above all, actions that target solutions to transform current land-use practice—agriculture and forestry—must be prioritized because they cut across all four crises.

To resolve this, bottom-up approaches are required to deliver the step change needed in the way land is used and greater efficiencies of forest commodity production. We know, however, that to achieve greater intensification of productivity, existing performance standards designed to manage linear incremental change will not suffice. Future standards must be designed to manage the complexity of systemic transformational changes.

As proposed by Rockström et al. (2017):

> … an explicit carbon roadmap for halving anthropogenic emissions every decade, codesigned by and for all industry sectors, could help promote disruptive, nonlinear technological advances toward a zero-emissions world. The key to such a roadmap will be a dual strategy that pushes innovative strategies up the creation and dissemination trajectory, while simultaneously pulling fossil-based value propositions from the market. Thus, the transformation unfolds at a pace governed by innovation rather than by inertia imposed by incumbent technologies.

While the forestry sector could clearly play a pivotal role in climate change mitigation and in such a carbon roadmap, the reality is that the forest sector has little influence in global thought leadership. In the context of the UN SDGs, forests are incorrectly framed in a narrow and restricted perspective limited to biodiversity (sustainable forest management is mentioned only once in SDG 13). The global discourse cannot be limited to the biodiversity area alone, otherwise the active management or restoration of forests will simply be invisible. A narrow interpretation of forests with most of their social and economic benefits and functions disregarded or unrecognized greatly limits options for a bioeconomy, climate resilience, decarbonization, and fossil fuel substitution and appropriate integration of their true socioeconomic value.

Part of the underlying hesitancy to political endorsement of forestry in the climate agenda could lie in the fact that such an endorsement would require massive adoption of plantation forestry and, in particular, IMPFs. Given the fact that genetically engineered (GM) trees will play a growing role in yield enhancement and protection and in modification of feedstock chemistry (Indufor 2012), it has been our experience that it becomes very problematic to even convene appropriate stakeholders for debate.

Nevertheless, a new narrative is needed that reframes global environmental change from an exclusive focus on physical processes of nature to the interlinkages with social processes. This new insight would provide an incentive to see the environment in terms of use and distribution of natural resources rather than scarcity. This subtle switch could have powerful consequences to enable a more rapid and inclusive uptake of practical measures to tackle production and consumption and internalize them into mainstream economic strategy.

Our view is that part of this narrative, based on our experience to date, must be the moral imperative of unlocking the potential of tree plantation biotechnology at scale, on time, and in the right location.

We believe that the landscape level experimentation we propose will contribute significantly to this new narrative, both through providing robust scientific evidence for decision making, but also by providing a platform for objective debate on the reality of what is possible.

The availability of a landscape level experiential platform for debate could help build a positive, mobilizing vision for the ways in which communities and biodiversity can benefit from the technologies that will underpin intensification and insights for appropriate governance frameworks. For both civil society and business, this means developing a cooperative model that favors a diffusion of advanced technologies to the benefit of more, not fewer, people under conditions that are acceptable. This would be unprecedented but necessary since we live in a world in which the only thing we can be certain of is change. Change challenges everything. It challenges most of all the tools we employ to achieve our objectives—if our objectives are to remain relevant and have impact at scale. The speed of change is the most difficult facet of this challenge because it requires us to act in ways and in contexts in which we are unfamiliar. Exponential change is inevitable, and the risks of not mounting an effective systemic response are profound. Collective response in developing transformative solutions will start with trust building to enable the collusion required for implementation. Governance will provide the cohesion required to create the resilience necessary for meaningful impact.

This demands an enormous upheaval in our institutional and personal behavior toward collusion on an unprecedented level. If joint interests can be agreed upon up front under appropriate governance frameworks, there is a more mobilizing potential for development and deployment of innovations that could enhance carbon drawdown and climate mitigation. Our interests are therefore roughly aligned and opposition to such technologies could actually do more harm than good to local communities.

Our GM tree policy was drafted from the perspective that it is in our interest to share innovation broadly with our outgrowers and to make it freely available for humanitarian and forest protection purposes. It need not be the case that the only way of balancing more with less through improving technical efficiency ("intensification") is seen only as an interim solution for resolving the imbalance between the activities of society and the capacities of the natural world. This could in fact be the start of a trajectory toward renewal.

We are also confident that the dialogue platform we propose could address the evident hesitancy in considering high-tech plantations as an option for development. Dialogue of such a nature could provide the reassurances and evidence of safeguards and shared benefits that managing transformational change is practically possible and that disruption need not necessarily be a threat.

For business this is a major challenge; it will involve more intimately working with nongovernmental organizations (NGOs), integrating "social license to operate" into strategy, diffusion rather than consolidation of innovation, and internalizing environmental externalities. The most fundamental change required would be the building of trust, to deconstruct ideological barriers that thwart collective action, and enable the identification of a common ground to achieve a common objective of sustainable forest management.

A recent publication describing a set of principles for governance of crop biotechnology (Gordon et al. 2021), significantly shifts the debate toward collective action and the question of shared value. This offers a novel path toward defining

common ground on which such principles could be implemented. "Risks and benefits should be weighed against the counterfactuals of continuing with business as usual against the backdrop of a changing climate and growing human population" (Gordon et al. 2021). In January 2022, in collaboration with members of the steering committee of this publication, we published a commentary on this article (May et al. 2022), because many of the key features of these principles concur broadly with the GM tree policy we had published earlier. We welcome a pathway for engagement and exploration of narratives toward further positive understanding of the relevance of biotechnology to climate change food security and related challenges.

The moral imperative should be to embrace newer methods, allowing advances to leave the laboratory and be tested in the field unhindered. This is of special importance for developing countries, where modern plant-breeding methods can allow local scientists to innovate and improve the crops needed to feed their citizens (May et al. 2022).

Ultimately, success lies not only in the science available, but in parallel, the co-creation of an environment in which science is embraced and capacity building and technology transfer flow from that. Science, without public acceptance, will in the end be irrelevant, and a huge, existential global imperative will be sidelined through misunderstanding, causing irreparable planetary damage.

The fundamental challenge and opportunity of our time is therefore to develop leadership in the formulation of a governance framework that will master natural resource productivity efficiency in transformative and socially relevant ways. Within this, the physical challenge is to develop and deploy the science and technology for the sustainable intensification of agroforestry commodity production. The socioeconomic challenge will be to ensure that technology reaches those who need it the most and that responsible procurement policies are systemic. The environmental challenge is to uncouple solutions for meeting productivity challenges from persistent cycles of ecosystem degradation and carbon emissions. Overarching all of this will be a behavioral challenge at all levels to understand and embrace scientific and technological innovation as a means for reducing complexity to practice via a meaningful global dialogue and convergence between the private sector, the scientific community, civil society, and governments.

If the challenges we face collectively are complex, then so are the solutions. The way forward is enhanced and informed dialogue on technology, enabling policies that stimulate the flows of technology and investment required and international cooperation in technology development and deployment. Well managed, and appropriately located GM plantations can be a key feature of healthy, diverse, and multifunctional forest landscapes, and can thus provide conditions that are compatible with climate, biodiversity conservation, and human well-being needs.

Current thought leadership identifies the need for positive exponentials to break the existing paradigm of economic, ecological, and social dysfunction (Elkington 2020). Many of the technologies and ideas already exist. The fundamental shift required to unleash that potential and drive a regenerative, distributive economic order is the creation of hitherto unimaginable coalitions that can transcend current ideological and institutional barriers and disrupt dysfunction. More specifically, the scale and pace of breakthrough will be determined by the speed with which the technologies and ideas of the future can gain social license.

ACKNOWLEDGMENTS

This article has been made freely available online by generous financial support from the Bill and Melinda Gates Foundation. We particularly want to acknowledge the support and encouragement of Rodger Voorhies, president, Global Growth and Opportunity Fund.

REFERENCES

*Reference is also in this subject collection.

Abramson M, Shoseyov O, Hirsch S, Shani Z. 2013. Genetic modification of plant cell walls to increase biomass and bioethanol production. In *Advanced biofuels and bioproducts* (ed. Lee J), pp. 315–338. Springer, New York.

Attenborough D. 2020. How to restore our forests. https://www.ourplanet.com/en/video/how-to-restore-our-forests

Betti M, Bauwe H, Busch FA, Fernie AR, Keech O, Levey M, Ort DR, Parry MA, Sage R, Timm S, et al. 2016. Manipulating photorespiration to increase plant productivity: recent advances and perspectives for crop improvement. *J Exp Bot* **67**: 2977–2988. doi:10.1093/jxb/erw076

Binkley D, Campoe OC, Alvares C, Carneiro RL, Cegatta I, Stape JL. 2017. The interactions of climate, spacing and genetics on clonal *Eucalyptus* plantations across Brazil and Uruguay. *For Ecol Manage* **405**: 271–283. doi:10.1016/j.foreco.2017.09.050

Coelho AJP, Villa PM, Matos FAR, Heringer G, Bueno ML, de Paula Almado R, Meira-Neto JAA. 2022. Atlantic forest recovery after long-term eucalyptus plantations: the role of zoochoric and shade-tolerant tree species on carbon stock. *For Ecol Manage* **503**: 119789. doi:10.1016/j.foreco.2021.119789

da Silva PHM, Poggiani F, Libardi PL, Gonçalves AN. 2013. Fertilizer management of eucalypt plantations on sandy soil in Brazil: initial growth and nutrient cycling. *For Ecol Manage* **301**: 67–78. doi:10.1016/j.foreco.2012.10.033

de Freitas CH, Bolzan Martins F. 2019. Thermal requirements and photoperiod influence in the leaf development of two forest species. *Floresta Ambient* doi:10.1590/2179-8087.001319

Elkington J. 2020. *Green swans: the coming boom in regenerative capitalism*, p. 284. Fast Company, New York.

Elli EF, Sentelhas PC, Bender FD. 2020. Impacts and uncertainties of climate change projections on *Eucalyptus* plantations productivity across Brazil. *For Ecol Manage* **474**: 118365. doi:10.1016/j.foreco.2020.118365

*Erb TJ. 2023. Photosynthesis 2.0: realizing new-to-nature CO_2-fixation to overcome the limits of natural metabolism. *Cold Spring Harb Perspect Biol* doi:10.1101/cshperspect.a041669

European Commission. 2020. 2050 long-term strategy. https://ec.europa.eu/clima/policies/strategies/2050_en

FAO. 2020. Global forest resources assessment 2020: main report. Food and Agriculture Organization of the United Nations, Rome. doi:10.4060/ca9825en

Florencio GWL, Bolzan Martins F, Azevedo Fagundes FF. 2022. Climate change on eucalyptus plantations and adaptive measures for sustainable forestry development across Brazil. *Ind Crops Prod* **188**: 115538. doi:10.1016/j.indcrop.2022.115538

Flores Carreño S, Macario C, Harel T. 2012. The roadmap for Vision 2050: an implementation guide for executing the Vision 2050 report. WBCSD, Geneva. http://docs.wbcsd.org/2012/09/Vision2050_Road_Map.pdf

FuturaGene. 2023. https://www.futuragene.com

Gordon DR, Jaffe G, Doane M, Glaser A, Gremillion TM, Ho MD. 2021. Responsible governance of gene editing in agriculture and the environment. *Nat Biotechnol* **39**: 1055–1057. doi:10.1038/s41587-021-01023-1

Grossberg SP. 2009. *Forest management*, p. 329. Nova, Hauppauge, NY.

IBÁ. 2022. Statistical data. Brazilian Tree Industry, Brasília, Brazil. https://iba.org/eng/statistical-data

Indufor. 2012. Strategic review on the future of forest plantations in the world. Study done for the Forest Stewardship Council (FSC). Bonn, Germany.

ITTO. 2020. Annual report. https://www.itto.int/direct/topics/topics_pdf_download/topics_id=6775&no=1

IUCN. 2009. Billion hectares of forests with potential for restoration, study shows. https://www.iucn.org/content/billion-hectares-forests-potential-restoration-study-shows

Jagels R. 2006. Management of wood properties in planted forests: a paradigm for global forest production. FAO working paper. https://www.fao.org/3/j8289e/J8289E00.htm

Kanowski P, Murray H. 2008. *Intensively managed planted forests: towards best practice.* The Forests Dialogue (TFD), New Haven, CT.

Martins FB, Bitencourt Benassi R, Rodriques Torres R, Agustinho de Brito Neto F. 2022. Impacts of 1.5°C and 2°C global warming on *Eucalyptus* plantations in South America. *Sci Total Environ* **825:** 153820. doi:10.1016/j.scitotenv.2022.153820

May M, Hirsch S. 2015. Commercialization of forestry genetic research: from promise to practice. In *Forests and globalization—challenges and opportunities for sustainable development* (ed. Nikolakis W, Innes J). Routledge, London.

May M, Giddings V, DeLisi C, Drell D, Patrinos A, Hirsch S, Roberts R. 2022. Constructive principles for gene editing oversight. *Nat Biotechnol* **40:** 17–18. doi:10.1038/s41587-021-01189-8

Nature Biotechnology. 2015. Brazil approves transgenic eucalyptus. *Nat Biotechnol* **33:** 577. doi:10.1038/nbt0615-577c

Ramantswana M, Guerra SPS, Ersson BT. 2020. Advances in the mechanization of regenerating plantation forests: a review. *Curr For Rep* **6:** 143–158. doi:10.1007/s40725-020-00114-7

Rockström J, Gaffney O, Rogelj J, Meinshausen M, Nakicenovic N, Schellnhuber HJ. 2017. A roadmap for rapid decarbonization. *Science* **355:** 1269–1271. doi:10.1126/science.aah3443

Scheffers BR, De Meester L, Bridge TCL, Hoffmann AA, Pandolfi JM, Corlett RT, Butchart SHM, Pearce-Kelly P, Kovacs KM, Dudgeon D, et al. 2016. The broad footprint of climate change from genes to biomes to people. *Science* **354:** aaf7671. doi:10.1126/science.aaf7671

Scolforo HF, McTague JP, Burkhart H, Roise J, Alcarde Alvares C, Stape JL. 2019. Modeling whole-stand survival in clonal eucalypt stands in Brazil as a function of water availability. *For Ecol Manage* **432:** 1002–1012. doi:10.1016/j.foreco.2018.10.044

Sembiring N, Napitupulu HL, Sipahutar AI, Sembiring MT. 2020. A review of sustainable replanting eucalyptus: higher sustainable productivity. *IOP Conf Ser: Mater Sci Eng* **935:** 012068. doi:10.1088/1757-899X/935/1/012068

Shani Z, Dekel M, Tsabary G, Goren R, Shoseyov O. 2004. Growth enhancement of transgenic poplar plants by over-expression of *Arabidopsis thaliana* endo-1,4-β-glucanase (*cel1*). *Mol Breed* **14:** 321–330. doi:10.1023/B:MOLB.0000049213.15952.8a

South PF, Cavanagh AP, Lopez-Calcagno PE, Raines CA, Ort DR. 2018. Optimizing photorespiration for improved crop productivity. *J Integr Plant Biol* **60:** 1217–1230. doi:10.1111/jipb.12709

Venkata BP, Polzin R, Wilkes R, Fearn A, Blumenthal D, Rohrbough S, Taylor NJ. 2020. Heterologous overexpression of *Arabidopsis cel1* enhances grain yield, biomass and early maturity in *Setaria viridis*. *Front Plant Sci* **11:** 515078. doi:10.3389/fpls.2020.515078

Wen YG, Zhou XG, Yu SF, Zhu HG. 2018. The predicament and countermeasures of development of global Eucalyptus plantations. *Guangxi Sci* **25:** 107–116.

World Economic Forum. 2021. COVID-19 is threatening the SDGs—here's what needs to happen. https://www.weforum.org/agenda/2021/07/sdgs-covid19-poverty-goals

World Resources Institute. 2021. State of climate action 2021: systems transformations required to limit global warming to 1.5°C. https://www.wri.org/research/state-climate-action-2021

World Wildlife Federation. 2011. The living forest report. https://wwf.panda.org

 Cite this article as *Cold Spring Harb Perspect Biol* doi: 10.1101/cshperspect.a041670

Overcoming Obstacles to Gene-Edited Solutions to Climate Challenges

L. Val Giddings

Information Technology and Innovation Foundation, Washington, D.C. 20001, USA

Correspondence: vgiddings@itif.org

Gene editing and genetic modification hold enormous potential to deliver solutions to multiple climate change challenges. The most important rate-limiting obstacles impeding their development and deployment are not technical, but rather counterproductive policies and regulations. These are driven in part by the mistaken apprehension of widespread public opposition. These obstacles are described and solutions to overcoming them are presented.

Jared Diamond observed that,

> Any society goes through social movements or fads, in which economically useless things become valued or useful things devalued temporarily. Nowadays, when almost all societies on Earth are connected to each other, we cannot imagine a fad's going so far that an important technology would actually be discarded. A society that temporarily turned against a powerful technology would continue to see it being used by neighboring societies and would have the opportunity to reacquire it by diffusion (or would be conquered by neighbors if it failed to do so).
>
> —Diamond (1997)

Crops improved through biotechnology, often called genetically modified organisms (GMOs),[1] have made major contributions to improved agricultural productivity and sustainability, increasing farmer incomes, improving environmental health, food safety, and benefitting consumers worldwide (Klümper and Qaim 2014; Nicolia et al. 2014; ISAAA 2020; Brookes 2022a,b).

Gene editing has already accelerated the rate at which such benefits are being imagined, developed, and delivered and holds the potential to help rapidly address some of the critical challenges associated with reducing greenhouse gas (GHG) emissions, as others have shown (Asanuma and Ozaki 2020; DeLisi et al. 2020; Giddings et al. 2020; Ito 2021; Houser 2022; Rosenzweig 2022; Zahoor 2022). Yet, despite these considerable benefits and the urgent need for these solutions, opposition and obstacles threaten to delay or block such beneficial applications. This paper describes the most important obstacles and shows how they can be overcome.

RATE-LIMITING FACTORS

Other papers have described several gene-edited innovations that hold significant promise for reducing GHG emissions and/or drawing down atmospheric levels of carbon dioxide. Technical

---

[1] Some have tried to draw a distinction between "genetic engineering" by which they mean the use of recombinant DNA techniques to insert exogenous DNA into a genome, and "gene editing" defined as limited to tweaking extant DNA sequences in a genome. But as technology advances, gene-editing techniques are increasingly being used to achieve similar or identical results as genetic engineering (Irving 2022; Yarnall et al. 2022) so we use the terms as more or less synonymous.

Cite this article as *Cold Spring Harb Perspect Biol* doi: 10.1101/cshperspect.a041677

obstacles remain to be conquered before these and other related innovations can be reduced to practice and deployed at scale but the promise for sizable and rapid impacts is high. The primary rate-limiting factor is not technical challenges, but regulatory and policy burdens based on the perception of problems with public acceptance (Lee 2022).

Public acceptance is difficult to measure. Opinion surveys of widely differing quality have delivered varying results over the past four decades. The common wisdom suggests there is considerable public hesitation, if not downright resistance, at least to foods improved through biotechnology. But while there are some regions in which this may be true—e.g., Austria—the reality is more complicated (Keller 2021), and the data tell a very different story.

There is often a gap, sometimes dramatic, between what people self-report about their attitudes in a public opinion survey, and their attitudes as manifest in actions. People often say one thing and do another. Indeed, it has been known for decades that many people simultaneously hold mutually contradictory views, particularly when it comes to genetic technologies (Office of Technology Assessment [OTA] United States Congress 1987). This is particularly so with regard to survey responses versus point-of-purchase decisions on whether or not to buy a product. But consumer data from around the world show consumers consistently base food purchase decisions on three factors: cost, taste, and quality. This holds true even among populations where a significant number claims otherwise (International Food Information Council [IFIC] 1921; Office of Technology Assessment [OTA] United States Congress 1987).

The reality is that every bushel of genetically engineered corn or soy that has been grown has been sold, and the market has not delivered a consistent premium for non-genetically engineered varieties. (Some soy varieties have consistently commanded premium prices for specialty varieties such as those used for tofu; but the market size for such varieties is too small to support the regulatory costs for approval of genetically modified [GM] varieties at present) (McDougall 2011; Lassoued et al. 2019; Whelan et al. 2020).

The agronomic and commercial success of crops improved through biotechnology has led directly to the explosive adoption of biotechnology-improved seed by farmers around the world (ISAAA 2020). This holds true not only for countries where biotech-improved seeds have been approved by governments, but also where farmers have been denied access by governments that have been slow to recognize the reality of their safety and superior sustainability (Giddings 2019). But the mistaken perception of broad consumer reluctance has been both a cause and effect of policies and practices that needlessly discriminate against biotechnology innovations in agriculture.

The result is policies, regulations, and business practices such as "non-GMO" certification, all of which create disincentives and barriers to the development and deployment of innovative solutions developed with gene editing and genetic engineering. This is the case around the world despite the lack of scientific justification for such discrimination, and in the face of massive experience demonstrating the superior safety and sustainability of such technologies (Smyth 2022).

The barriers created by these regulations, policies, and practices must be overcome if products of gene editing/genetic engineering are to be developed and deployed in a timely manner.

WHAT IS TO BE DONE? 1—REDUCING REGULATORY BURDENS

There is no doubt that well-designed regulations have contributed to enormous societal benefits by improving the safety of foods, devices, and tools, and reducing the unintended negative consequences of innovations (Havens 1978; Guasch and Hahn 1999; O'Toole 2014). But regulations that are not well designed and impose compliance burdens disproportionate to the benefits they aspire to deliver are counterproductive, and not "fit for purpose."

In the name of safety, regulatory regimes around the world impose burdens on the introduction and use in food and agriculture of innovations developed through biotechnology. These burdens are disproportionate to the actual hazards the innovations embody. The sanitary and

Cite this article as *Cold Spring Harb Perspect Biol* doi: 10.1101/cshperspect.a041677

phytosanitary (SPS) agreement of the World Trade Organization (WTO), and the International Plant Protection Convention (IPPC) codify standards that safety regulations must meet to ensure regulations actually advance safety rather than provide a cover for surreptitious goals, such as protectionism. Distilled to their essence, these standards stipulate that regulatory measures must address an actual threat (i.e., exposure to an identified hazard), not a hypothetical fear; they must be proportional to that threat; and they must not be excessive, more restrictive than necessary to manage or mitigate the threat. These standards form the foundation of the rules-based system of international trade in food commodities, on which food security for much of the world depends. But regulations applied to crops and foods improved through biotechnology around the world fall short of complying with these standards.

Governments generally take one or the other of two different approaches to safety assurance for crops and foods improved through biotechnology: one aims to ground regulations and approval decisions in data and build on experience; the other, despite claiming otherwise, elevates other criteria, such as "precaution" or political concerns. The United States, Australia, Canada, Japan, Argentina, Brazil, and some others fall into the first category, although not perfectly. The European Union and countries that follow the EU example, particularly in tropical latitudes, fall into the latter category.

The United States first promulgated a major biotechnology policy statement in 1986, publishing an approach to regulation known as the Coordinated Framework (CF) (Office of Science and Technology Policy [OSTP] Executive Office of the President 1986). The CF is based on the recognition that no novel risks of GMOs have been identified (i.e., no potential for exposure to novel hazards) compared to those familiar with crops and foods developed with traditional methods of genetic improvement (e.g., classical breeding, wide crosses, tissue culture) (National Academy of Sciences 1987, 1989; Kuiper et al. 2001; European Commission 2010). The OSTP, therefore, posited that existing legislation assigning authority to manage and mitigate such famil-

iar hazards to avoid unreasonable risks should be adequate to the task. Public comment was solicited and received confirming these facts. Regulatory agencies within the U.S. Department of Agriculture, Animal and Plant Health Inspection Service (USDA), Environmental Protection Agency (EPA), and Food and Drug Administration (FDA) were tasked with developing regulations, as necessary, to implement the CF, which they did in accord with their own authorities.

The FDA determined that existing methods for ensuring food safety were sufficient—sellers are responsible for the safety of the food they sell, and if a food had a history of safe consumption, it would avoid major premarket regulatory scrutiny. If the food was novel in some potentially hazardous way and lacked a history of safe use, the FDA laid out a consultation mechanism in which the agency asked questions to test the basis for a seller's claim of its safety.

The EPA developed a series of policies to confirm the safety of new pesticidal substances developed through biotechnology, but the USDA was the first agency to promulgate new regulations to deal with "GMOs." In accordance with the standards laid out in the CF, the U.S. Department of Agriculture, Animal and Plant Health Inspection Service (USDA) (1987) defined the regulatory trigger capturing "GMOs" for review according to the characteristics of the product—whether or not it presented a potential plant pest risk. This was established by the presence or absence of DNA sequences from a listed plant pest—the mere use of recombinant DNA technologies is/was insufficient to trigger regulatory oversight. Experience soon showed this criterion was not, in fact, a reliable marker for hazard (National Academy of Sciences 1989, 2000, 2002; Organization for Economic Cooperation and Development [OECD] 1992, 1993), and the Animal and Plant Health Inspection Service (APHIS) revised its application of the trigger to narrow the scope of items captured for premarket review and streamline the process (U.S. Department of Agriculture, Animal and Plant Health Inspection Service [USDA] 1992).

According to the CF, this approach of gaining experience with classes of new products (e.g., herbicide-tolerant maize, insect-resistant cotton, confirming safety, and reducing regulatory burdens while refocusing attention on areas of remaining uncertainty) should have become a regular pattern. Despite some encouraging moves in recent years, this has not happened (U.S. Department of Agriculture 2018). And over the past four decades, all U.S. regulatory agencies have drifted from the path envisioned in the CF, although none have wandered farther than the FDA. In its most recent proposals to ensure the safety of plants and animals improved through gene editing, the FDA has adopted a "guilty until proven innocent" approach that flies in the face of reason and experience (Van Eenennaam and Young 2014; Giddings 2017a,b). Not only is this approach not what is required under the law, but it is also scientifically nonsensical (Van Eenennaam et al. 2019; Van Eenennaam 2022).

As bad as the situation has been in the United States, it is far worse elsewhere in the world. The European Union adopted a "precautionary approach" that presumes biotech-improved seeds present novel hazards and unreasonable risks despite the absence of supporting data or experience (European Commission 1990; Paarlberg 2001; Entine 2006). The EU instituted a process so burdensome and political as to block nearly all farmer adoption of seeds improved through biotechnology and causing the world leading seed innovation industry, once found in Europe, largely to relocate to the United States (Torry 2012). Innovation in European agriculture has thus lagged behind the rest of the industrial world, as has also been the case for other countries that have followed the European example (ISAAA 2020).

The global situation today is that in virtually every jurisdiction, the disharmony between the degree of regulatory scrutiny applied to biotechnology-derived innovations in food and agriculture, and the level that would be proportional to the actual hazards and risks, has grown from a gap to a chasm. This degree of regulatory scrutiny does nothing to add to citizen safety, and such regulations are not "fit for purpose" (United Kingdom Advisory Committee on Releases to the Environment [UKACRE] 2013a,b, c; Gould et al. 2022). They serve only to prolong

reliance on older, if not obsolete, technologies, the products of which are generally less safe and less efficient than more recent innovations.

This means that the use of the most advanced seed-improvement technologies, with a remarkable record of safety and improvements in sustainability and productivity, are disincentivized and discriminated against despite such policies being contradicted by vast amounts of data and experience. Inasmuch as genomes throughout nature are salted with sequences imported from other lineages (Ridley 1999; Giddings 2015) to the extent that everything that appears on a dinner plate throughout most of the world is naturally transgenic, the irony is palpable. This represents a massive failure of policies in countries around the world that is especially egregious in the face of present challenges and needs. The opportunity costs of such retrograde policies are considerable (Gouse et al. 2016; McFadden et al. 2021; Usla 2022; Paarlberg and Smyth 2023).

The solution is as simple as it is obvious: regulations lacking justification in data and experience must be set aside as rapidly as possible. Detailed accounts of how this can/should be accomplished are not lacking (United Kingdom Advisory Committee on Releases to the Environment [UKACRE] 2013a,b,c; European Academies Science Advisory Council [EASAC] 2013; Conko et al. 2016; Giddings 2017a, 2021, 2022; Eriksson 2018; Gould et al. 2022). Their recommendations should be taken up and acted upon as a matter of urgency in the United States, Europe, and other countries around the world.

WHAT IS TO BE DONE? 2—ELIMINATING MISLEADING AND FRAUDULENT LABELS

Unduly burdensome regulations and innovation-hostile policies are the major impediments to the development and deployment of biotechnology-derived solutions to climate change problems and other societal challenges. But other forces also play an inimical role, including ill-founded labeling requirements or standards.

Some governments, particularly those in the European Union and its emulators, under the rubric of informing consumers, have mandated

Cite this article as *Cold Spring Harb Perspect Biol* doi: 10.1101/cshperspect.a041677

labels on foods containing "GMO"-derived ingredients above a certain threshold. The criteria triggering the EU "GMO" labeling requirements use the following language (European Commission 1997):

- Foods and food ingredients containing or consisting of GM organisms within the meaning of Directive 90/220/EEC;

- Foods and food ingredients produced [sic] from, but not containing, genetically modified organisms.

The definition of "GMO" found in Directive 90/220 is "genetically modified organism (GMO) means an organism in which the genetic material has been altered in a way that does not occur naturally by mating and/or natural recombination" (European Commission 1990).

The problems with this approach are manifold. First, the triggering content threshold of 0.9% is purely arbitrary, and bears no relationship to any meaningful measure of health, safety, nutrition, quality, etc. Second, the class of foods and food ingredients captured for labeling contains multiple exceptions, also arbitrary and unrelated to any meaningful criterion related to health, safety, nutrition, or otherwise (European Parliament 2003). And finally, and fatally, the definition of GMO as "an organism in which the genetic material has been altered in a way that does not occur naturally by mating and/or natural recombination" is purely nonsensical. Every "genetic modification" made in the production of any GMO or gene-edited product is made with enzymes scientists first found in nature; they are used to effect the same kinds of genetic changes observed widely throughout nature. Examples of discovery of gene transfer found in nature that are directly analogous, if not entirely indistinguishable from genetic modifications effected by scientists using naturally occurring enzymes to effect naturally occurring mechanisms of genetic change in the laboratory, are abundant (Wang et al. 2006; Schneider and Thomas 2014; Yong 2014, 2017; Gasmi et al. 2015; Jones 2015; Kyndt et al. 2015; Matveeva and Otten 2019; Chen et al. 2022). Far from de-

fining any meaningful class, this definition of "GMO" does little more than signal the biological illiteracy of those who conjured it (Johnson 2015; Tagliabue 2015, 2016, 2020; Wood 2022). Together with requirements for traceability, labeling, and testing, the European Union has constructed a compliance nightmare for which there is no shred of scientific justification, in gross violation of its obligations and responsibilities under the SPS/WTO and the IPPC (World Trade Organization 1994; IPPC 1997).

The argument that consumers want and have a "right to know" is equally unsound. Consumers fed a steady diet of propaganda and lies for years by vested interests (Academics Review 2014; Wetaya 2022b) without meaningful pushback from the governments supposedly safeguarding their interests will, of course, say they want labels to tell them whether or not they are buying GMOs, even though nobody can give them a definition of GMO that is not nonsense. But experience shows consumers reliably respond in the affirmative whenever they are asked if they want to have information on a label or otherwise available, because they are rightly skeptical of those who would propose to withhold meaningful information (Hackleman 1981). But when presented with a blank slate asking them what information they want on a label, very few indicate they want information on "GMOs." Consumer demand for GMO labels, in this case, is an artifact of inept opinion polls and ongoing propaganda campaigns by vested interests (Academics Review 2014; English 2019; Food Insight 2021; Lynas et al. 2022).

Perhaps the most egregiously false and misleading GMO label is not mandated by the government, but results from a business plan that appears to be based on fomenting unfounded fears about competing products and misleading consumers. The Non-GMO Project, a company created by a coalition of organic marketers will, for a fee, license the use of their label (a butterfly logo) to deceive consumers through false and misleading claims about foods, food ingredients, and their health and safety characteristics (Savage 2016; Giddings and Atkinson 2018; Miller 2018). Their marketing uses the label as a symbol to reassure consumers that a product is free of GMOs, even though they admit on their own website that it in fact provides no such guarantee. Such misleading labels defy both FDA and Canadian guidance on the topic and are in clear violation of the law (Campbell 2018; English 2019; FDA 2019; Canadian General Standards Board 2021). Such propaganda campaigns from vested interests contribute to the demonization of safe, more sustainable products and valuable technologies. Tolerating them imposes significant societal costs (Gelski 2016; English 2022).

WHAT IS TO BE DONE? 3—ACCELERATING PUBLIC ACCEPTANCE IS THE ESSENTIAL PREREQUISITE

Improving the acceptance climate among policymakers and regulators is essential if solutions developed through genetic engineering and gene editing are to be deployed. Many different actions can help improve the acceptance climate, from increased participation by scientists and informed citizens in science communication and policymaking, to sustained programs by credible independent entities including NGOs and think tanks.

Pushing back against the concerted disinformation campaigns from special interests that have driven such discriminatory policies is difficult, particularly for governments, but independent, science-based voices are well suited for the task. There are several entities with proven track records in this space, and publications from several are cited in the references below:

- The Genetic Literacy Project (GLP): https:// geneticliteracyproject.org;
- The Institute for Food and Agricultural Literacy (IFAL): https://ifal.ucdavis.edu;
- The International Service for the Acquisition of Agri-Biotech Applications (ISAAA): https:// www.isaaa.org/kc/default.asp;
- PG Economics: https://pgeconomics.co.uk/ publications; and the
- Information Technology and Innovation Foundation (ITIF): https://itif.org/issues/agricultural-biotech.

These groups are critically important contributors to enabling future innovations built on genetic technologies.

There are reasons to be hopeful that these obstacles to gene-edited and genetically engineered solutions will be overcome (Baksi 2022; Bayer 2022; Bounds and Terazono 2022). As noted above, the existing regulatory regimes have been harshly criticized, particularly in recent years. And decades of empty alarms of health and environmental calamities from the use of GMOs have corroded the credibility of opposition groups (Brown 2022; Gelski 2022), even as some of them have changed course (Morgan 2021; RePlanet 2022). Governments, particularly those in developing countries, have noticed the considerable benefits delivered by crops improved through biotechnology and are moving to avail themselves of the same rewards (Awal 2022; Das 2022; Maina 2022; Mong'ina 2022; NatureNews 2022; Wetaya 2022a; Zongo 2022). The perception of consumer opposition also seems to be declining (Food Insight 2021; Baksi 2022; Bayer 2022). But the need for climate change solutions these technologies can deliver is urgent, and it would be imprudent to rely on the natural corrections Diamond (1997) identified to remove the remaining obstacles (Bayer 2022; Doudna 2022; The Economist 2022). Philanthropic support for the independent advocates identified above can make a critical impact and provide a rapid return on investment. Such support should be provided as rapidly as possible. The risks of action need to be compared not to a presumed no-risk-if-we-don't-act path, but to the very real, well understood, scientifically grounded, and truly dire risks of staying the current course.

ACKNOWLEDGMENTS

This article has been made freely available online by generous financial support from the Bill and Melinda Gates Foundation. We particularly want to acknowledge the support and encouragement of Rodger Voorhies, president, Global Growth and Opportunity Fund.

Cite this article as *Cold Spring Harb Perspect Biol* doi: 10.1101/cshperspect.a041677

REFERENCES

Academics Review. 2014. Why consumers pay more for organic foods? Fear sells and marketers know it. https://academics-review.bonuseventus.org/2014/04/why-consumers-pay-more-for-organic-foods-fear-sells-and-marketers-know-it

Asanuma N, Ozaki T. 2020. Japan approves gene-edited "uper tomato." But will anyone eat it? https://asia.nikkei.com/Business/Science/Japan-approves-gene-edited-super-tomato.-But-will-anyone-eat-it

Awal M. 2022. Nat'l Biosafety Authority approves first GM crop; says it will be of significant benefit. https://thebftonline.com/2022/08/23/natl-biosafety-authority-approves-first-gm-crop-says-it-will-be-of-significant-benefit

Baksi S. 2022. Environmental release of GM-mustard in India: cause for hope. https://fas.org.in/environmental-release-of-gm-mustard-in-india-cause-for-hope

Bayer E. 2022. Get ready for an explosion of productivity in biotechnology. https://www.forbes.com/sites/ebenbayer/2022/07/29/get-ready-for-an-explosion-of-productivity-in-biotechnology/?sh=eec6e6f1a945

Bounds A, Terazono E. 2022. Mood shifts on gene-edited crops as droughts and wars bite. https://www.ft.com/content/337ba132-cc33-42a9-9df3-cd53114200bc

Brookes G. 2022a. Farm income and production impacts from the use of genetically modified (GM) crop technology 1996–2020. *GM Crops Food* **13:** 171–195. doi:10.1080/21645698.2022.2105626

Brookes G. 2022b. Genetically modified (GM) crop use 1996–2020: environmental impacts associated with pesticide use change. *GM Crops Food* **13:** 262–289. doi:10.1080/21645698.2022.2118497

Brown W. 2022. Why genetically-modified crops may be about to have their day. https://www.telegraph.co.uk/global-health/climate-and-people/why-genetically-modified-crops-may-have-day

Campbell A. 2018. CFIA now says Non-GMO Project Verified doesn't mean non-GMO. https://www.realagriculture.com/2018/04/cfia-now-says-non-gmo-project-verified-doesnt-mean-non-gmo

Canadian General Standards Board. 2021. Voluntary labelling and advertising of foods that are and are not products of genetic engineering. CAN/CGSB 32.315-2004. Standards Council of Canada, Ottawa, Canada.

Chen K, Zhurbenko P, Danilov L, Matveeva T, Otten L. 2022. Conservation of an *Agrobacterium* cT-DNA insert in *Camellia* section *Thea* reveals the ancient origin of tea plants from a genetically modified ancestor. *Front Plant Sci* **13:** 997762. doi:10.3389/fpls.2022.997762

Conko G, Kershen DL, Miller H, Parrott WA. 2016. A risk-based approach to the regulation of genetically engineered organisms. *Nat Biotechnol* **34:** 493–503. doi:10.1038/nbt.3568

Das KN. 2022. India says GM technology important for food security, import reduction. https://www.reuters.com/world/india/india-says-gm-technology-important-food-security-import-reduction-2022-12-08

DeLisi C, Patrinos A, MacCracken M, Drell D, Annas G, Arkin A, Church G, Cook-Deegan R, Jacoby H, Lidstrom M, et al. 2020. The role of synthetic biology in atmospheric greenhouse gas reduction: prospects and challenges. *BioDesign Res* **2020:** 1–8. doi:10.34133/2020/1016207

Diamond J. 1997. *Guns, germs, and steel: the fates of human societies.* W.W. Norton, New York.

Doudna JA. 2022. Starting a revolution isn't enough. https://www.theatlantic.com/science/archive/2022/09/crispr-cas9-gene-editing-biotechnology/671382

English C. 2019. 30,000 food products with Non-GMO Project label may be "false or misleading," FDA guidance document says. https://geneticliteracyproject.org/2019/04/11/30000-food-products-with-non-gmo-project-label-may-be-false-or-misleading-fda-guidance-document-says

English C. 2022. Cost of ignoring science: non-GMO labels lead to lawsuits. https://www.acsh.org/news/2022/01/31/cost-ignoring-science-non-gmo-labels-lead-lawsuits-16091

Entine J. 2006. *Let them eat precaution: how politics is undermining the genetic revolution in agriculture.* American Enterprise Institute, Washington, DC.

Eriksson D. 2018. Recovering the original intentions of risk assessment and management of genetically modified organisms in the European union. *Front Bioeng Biotechnol* **6:** 52. doi:10.3389/fbioe.2018.00052

European Academies Science Advisory Council (EASAC). 2013. *Planting the future: opportunities and challenges for using crop genetic improvement technologies for sustainable agriculture.* German National Academy of Sciences Leopoldina, Halle, Germany.

European Commission. 1990. Council directive 90/220/EEC of 23 April 1990 on the deliberate release into the environment of genetically modified organisms. *Off J L* **117:** 15–27.

European Commission. 1997. Regulation (EC) No 258/97 of the European Parliament and of the Council of 27 January 1997 concerning novel foods and novel food ingredients. *Off J L* **43:** 1–6.

European Commission. 2010. *A decade of EU-funded GMO research (2001–2010).* European Union, Luxembourg.

European Parliament. 2003. Regulation (EC) No 1829/2003 of the European Parliament and of the Council of 22 September 2003 on genetically modified food and feed. *Off J L* **286:** 1–23.

FDA. 2019. Voluntary labeling indicating whether foods have or have not been derived from genetically engineered plants: guidance for industry. https://www.fda.gov/media/120958/download

Food Insight. 2021. International Food Information Council (IFIC) survey: knowledge, understanding and use of front-of-pack labeling in food and beverage decisions: insights from shoppers in the U.S. https://foodinsight.org/ific-survey-fop-labeling

Gasmi L, Boulain H, Gauthier J, Hua-Van A, Musset K, Jakubowska AK, Aury JM, Volkoff AN, Huguet E, Herrero S, et al. 2015. Recurrent domestication by Lepidoptera of genes from their parasites mediated by Bracoviruses. *PLoS Genet* **11:** e1005470. doi:10.1371/journal.pgen.1005470

Gelski J. 2016. G.M.O. labeling alone may cost Americans $3.8 billion. https://www.foodbusinessnews.net/articles/7433-g-m-o-labeling-alone-may-cost-americans-3-8-billion

Gelski J. 2022. Gen Z could respond favorably to GMOs. https://www.foodbusinessnews.net/articles/22693-gen-z-could-respond-favorably-to-gmos

Giddings LV. 2015. Hey, Chipotle! Don't look now, but it's "turtles all the way down!" https://itif.org/publications/2015/04/30/hey-chipotle-don%E2%80%99t-look-now-it%E2%80%99s-turtles-all-way-down

Giddings LV. 2017a. Comments to FDA on proposal to regulate gene-edited animals. https://itif.org/publications/2017/04/17/comments-fda-proposal-regulate-gene-edited-animals

Giddings LV. 2017b. Comments to Food and Drug Administration on gene-edited plants. https://itif.org/publications/2017/06/14/comments-food-and-drug-administration-gene-edited-plants

Giddings LV. 2019. Indian farmers launch civil disobedience campaign to secure access to GM seeds. https://itif.org/publications/2019/06/14/indian-farmers-launch-civil-disobedience-campaign-secure-access-gm-seeds

Giddings LV. 2021. How the Biden administration can accelerate prosperity by fixing agricultural-biotech regulations. https://itif.org/publications/2021/03/31/how-biden-administration-can-accelerate-prosperity-fixing-agricultural

Giddings LV. 2022. Comments to the European commission regarding regulation of plants produced with "New Genomic Techniques." The Innovation Files, July 6. https://itif.org/publications/2022/07/06/comments-to-the-european-commission-regarding-ngt-plant-regulation

Giddings LV, Atkinson RD. 2018. ITIF files FDA citizen petition to prohibit "non-GMO" labels. https://itif.org/publications/2018/09/24/itif-files-fda-citizen-petition-prohibit-non-gmo-labels

Giddings LV, Rozansky R, Hart DM. 2020. Gene editing for the climate: biological solutions for curbing greenhouse emissions. http://www2.itif.org/2020-gene-edited-climate-solutions.pdf

Gould F, Amasino RM, Brossard D, Buell CR, Dixon RA, Falck-Zepeda JB, Gallo MA, Giller KE, Glenna LL, Griffin T, et al. 2022. Toward product-based regulation of crops: current process-based approaches to regulation are no longer fit for purpose. Science 377: 1051–1053. doi:10.1126/science.abo3034

Gouse M, Sengupta D, Zambrano P, Zepeda JF. 2016. Genetically modified maize: less drudgery for her, more maize for him? Evidence from smallholder maize farmers in South Africa. World Dev 83: 27–38. doi:10.1016/j.worlddev.2016.03.008

Guasch JL, Hahn RW. 1999. The costs and benefits of regulation: implications for developing countries. World Bank Res Obs 14: 137–158. doi:10.1093/wbro/14.1.137

Hackleman EC. 1981. Food label information: what consumers say they want and what they need. Adv Consum Res 8: 477–483.

Havens HS. 1978. Costs and benefits of government regulation. https://www.gao.gov/assets/107970.pdf

Houser K. 2022. 7 ways CRISPR is shaping the future of food. https://www.freethink.com/science/crispr-food

IPPC. 1997. International plant protection convention (new revised text). https://www.ippc.int/en/publications/131

Irving M. 2022. New CRISPR gene-editing system can "drag-and-drop" DNA in bulk. New Atlas, https://newatlas.com/biology/paste-crispr-gene-editing-tool

ISAAA. 2020. Biotech crops drive socio-economic development and sustainable environment in the new frontier. https://www.isaaa.org/resources/publications/briefs/55/executivesummary/default.asp

Ito J. 2021. SCIENCEA new era for food: the potential benefits of gene-edited foods. https://japan-forward.com/a-new-era-for-food-the-potential-benefits-of-gene-edited-foods

Johnson N. 2015. It's practically impossible to define "GMOs." https://grist.org/food/mind-bomb-its-practically-impossible-to-define-gmos

Jones J. 2015. Domestication: sweet! a naturally transgenic crop. Nat Plants 1: 15077. doi:10.1038/nplants.2015.77

Keller D. 2021. Why does Europe oppose GMOs? https://www.dtnpf.com/agriculture/web/ag/crops/article/2021/03/02/europe-oppose-gmos

Klümper W, Qaim M. 2014. A meta-analysis of the impacts of genetically modified crops. PLoS ONE 9: e111629. doi:10.1371/journal.pone.0111629

Kuiper HA, Kleter GA, Noteborn HP, Kok EJ. 2001. Assessment of the food safety issues related to genetically modified foods. Plant J 27: 503–528. doi:10.1046/j.1365-313x.2001.01119.x

Kyndt T, Quispe D, Zhai H, Jarret R, Ghislain M, Liu Q, Gheysen G, Kreuze JF. 2015. The genome of cultivated sweet potato contains Agrobacterium T-DNAs with expressed genes: an example of a naturally transgenic food crop. Proc Natl Acad Sci 112: 5844–5849. doi:10.1073/pnas.1419685112

Lassoued R, Phillips PWB, Smyth SJ, Hesseln H. 2019. Estimating the cost of regulating genome edited crops: expert judgment and overconfidence. GM Crops Food 10: 44–62. doi:10.1080/21645698.2019.1612699

Lee C. 2022. Despite progress, GMO fears limit new traits, products. https://naturalresourcereport.com/2022/09/despite-progress-gmo-fears-limit-new-traits-products

Lynas M, Adams J, Conrow J. 2022. Misinformation in the media: global coverage of GMOs 2019-2021. GM Crops Food doi:10.1080/21645698.2022.2140568

Maina J. 2022. GM crops safe and viable option for Africa, say African science academy experts. https://allianceforscience.org/blog/2022/12/gm-crops-safe-and-viable-option-for-africa-say-african-science-academy-experts

Matveeva TV, Otten L. 2019. Widespread occurrence of natural genetic transformation of plants by Agrobacterium. Plant Mol Biol 101: 415–437. doi:10.1007/s11103-019-00913-y

McDougall P. 2011. The cost and time involved in the discovery, development and authorisation of a new plant biotechnology derived trait. Consultancy study for Crop Life International, Midlothian, pp. 1–24.

McFadden B, Van Eenennaam A, Goddard E, Lusk J, McCluskey J, Smyth, SJ, Taheripour F, Tyner WE. 2021. Gains foregone by going GMO free: potential impacts on consumers, the environment, and agricultural producers. https://animalbiotech.ucdavis.edu/sites/g/files/dgvnsk501/files/inline-files/QTA2021-2-GMO-Free-CAST.pdf

Cite this article as *Cold Spring Harb Perspect Biol* doi: 10.1101/cshperspect.a041677

Miller HI. 2018. The organic industry is lying to you. https:// www.wsj.com/articles/the-organic-industry-is-lying-to-you-1533496699?mod=djemMER

Mong'ina R. 2022. Gene editing viable solution to double crop production amidst food insecurity in Africa. https://shahi dinews.co.ke/2022/10/04/gene-editing-viable-solution-to-double-crop-production-amidst-food-insecurity-in-africa

Morgan K. 2021. The demise and potential revival of the American chestnut. https://www.sierraclub.org/sierra/ 2021-2-march-april/feature/demise-and-potential-revival-american-chestnut

National Academy of Sciences. 1987. *Introduction of recombinant DNA engineered organisms into the environment: key issues.* National Academies Press, Washington, DC.

National Academy of Sciences. 1989. *Field testing genetically modified organisms: framework for decisions.* National Academies Press, Washington, DC.

National Academy of Sciences. 2000. *Genetically modified pest-protected plants: science and regulation.* National Academies Press, Washington, DC.

National Academy of Sciences. 2002. *The environmental effects of transgenic plants: the scope and adequacy of regulation.* National Academies Press, Washington, DC.

NatureNews. 2022. Mozambique, Ethiopia benefit from Nigeria's achievements in biotechnology. https://sustaina bleeconomyng.com/mozambique-ethiopia-benefit-from-nigerias-achievements-in-biotechnology

Nicolia A, Manzo A, Veronesi F, Rosellini D. 2014. An overview of the last 10 years of genetically engineered crop safety research. *Crit Rev Biotechnol* **34:** 77–88. doi:10 .3109/07388551.2013.823595

Office of Science and Technology Policy (OSTP) Executive Office of the President. 1986. Coordinated framework for regulation of biotechnology; announcement of policy and notice for public comment. 51 FR 23302, Washington, DC.

Office of Technology Assessment (OTA) United States Congress. 1987. *New developments in biotechnology 2: public perceptions of biotechnology. OTA-BP-BA—45.* U.S. Government Printing Office, Washington, DC

Organization for Economic Cooperation and Development (OECD). 1992. *Safety considerations for biotechnology.* OECD, Paris.

Organization for Economic Cooperation and Development (OECD). 1993. *Biotechnology, agriculture and food.* OECD, Paris.

O'Toole J. 2014. The hidden business benefits of regulation. https://www.strategy-business.com/blog/The-Hidden-Business-Benefits-of-Regulation

Paarlberg RL. 2001. *The politics of precaution—genetically modified crops in developing countries.* International Food Policy Research Institute/Johns Hopkins University Press, Baltimore, MD.

Paarlberg R, Smyth SJ. 2023. The cost of not adopting new agricultural food biotechnologies. *Trends Biotechnol* **41:** 304–306. doi:10.1016/j.tibtech.2022.09.006

RePlanet. 2022. Finland's Green Party now supports genetic engineering in agriculture. https://www.replanet.ngo/ post/finland-s-green-party-now-supports-genetic-enginee ring-in-agriculture

Rosenzweig LF. 2022. Purple tomato is first genetically engineered plant to be deregulated through USDA's new regulatory status review process. https://lifesciences.mofo .com/topics/purple-tomato-is-first-genetically-engineered-plant-to-be-deregulated-through-usda-s-new-regulatory-status-review-process

Ridley M. 1999. *Genome: the autobiography of a species in 23 chapters.* Harper Collins, New York.

Savage S. 2016. The non-GMO food label is a lie. https:// www.forbes.com/sites/stevensavage/2016/06/11/the-non-gmo-food-label-is-a-lie/?sh=6a3f43034b70

Schneider SE, Thomas JH. 2014. Accidental genetic engineers. horizontal sequence transfer from parasitoid wasps to their lepidopteran hosts. *PLoS ONE* **9:** e109446. doi:10 .1371/journal.pone.0109446

Smyth SJ. 2022. Government decisions that needlessly extend food insecurity. https://saifood.ca/decisions-food-insecurity

Tagliabue G. 2015. The nonsensical GMO pseudo-category and a precautionary rabbit hole. *Nat Biotechnol* **33:** 907–908. doi:10.1038/nbt.3333

Tagliabue G. 2016. The meaningless pseudo-category of "GMOs." *EMBO Rep* **17:** 10–13. doi:10.15252/embr .201541385

Tagliabue GM. 2020. Some potato stories: a multiple tale of "anti-GMO" detrimental nonsense. https://www.europe anscientist.com/en/features/some-potato-stories-a-multi ple-tale-of-anti-gmo-detrimental-nonsense

The Economist. 2022. Science has made a new genetic revolution possible. https://www.economist.com/leaders/ 2022/08/25/science-has-made-a-new-genetic-revolution-possible

Torry H. 2012. European hostility pushes BASF plant biotech to US. https://www.marketwatch.com/story/european-hostility-pushes-basf-plant-biotech-to-us-2012-01-16

United Kingdom Advisory Committee on Releases to the Environment (UKACRE). 2013a. Genetically modified organisms: review of current EU regulations. https:// www.gov.uk/government/publications/genetically-modifi ed-organisms-review-of-current-eu-regulations

United Kingdom Advisory Committee on Releases to the Environment (UKACRE). 2013b. Genetically modified organisms: the case for new regulations. https://www.gov .uk/government/publications/genetically-modified-organi sms-the-case-for-new-regulations

United Kingdom Advisory Committee on Releases to the Environment (UKACRE). 2013c. Genetically modified organisms: improving risk assessments. https://www.gov .uk/government/publications/genetically-modified-organ isms-improving-risk-assessments

U.S. Department of Agriculture. 2018. Secretary Perdue issues USDA statement on plant breeding innovation. https://www.usda.gov/media/press-releases/2018/03/28/ secretary-perdue-issues-usda-statement-plant-breeding-innovation

U.S. Department of Agriculture Animal and Plant Health Inspection Service (USDA). 1987. Introduction of organisms and products altered or produced through genetic engineering which are plant pests or which there is reason to believe may be plant pests. 7 CFR 340, 52 Fed. Reg. 22,892. Washington, DC.

U.S. Department of Agriculture Animal and Plant Health Inspection Service (USDA). 1992. Genetically engineered organisms and products: notification procedures for the introduction of certain regulated articles; and petition for nonregulated status. 7 CFR 340, 57 Fed. Reg. 53,036. Washington, DC.

Usla H. 2022. Prohibición del maíz transgénico borraría 42% del PIB agro: CNA [Prohibition of transgenic corn would erase 42% of agricultural GDP: CNA]. https://www-elfinanciero-com-mx.translate.goog/economia/2022/11/24/prohibicion-de-maiz-trans-borraria-42-de-pib-agro-cna/?_x_tr_sl=auto&_x_tr_tl=en&_x_tr_hl=en&_x_tr_pto=wapp

Van Eenennaam AL. 2022. DNA is NOT a drug. And regulating genome edited research animals as a drug is unworkable. https://biobeef.faculty.ucdavis.edu/2022/03/09/dna-is-not-a-drug-and-regulating-genome-edited-research-animals-as-a-drug-is-unworkable

Van Eenennaam AL, Young AE. 2014. Prevalence and impacts of genetically engineered feedstuffs on livestock populations. *J Anim Sci* **92**: 4255–4278. doi:10.2527/jas.2014-8124

Van Eenennaam AL, Wells KD, Murray JD. 2019. Proposed U.S. regulation of gene-edited food animals is not fit for purpose. *NPJ Sci Food* **3**: 3. doi:10.1038/s41538-019-0035-y

Wang W, Zheng H, Fan C, Li J, Shi J, Cai Z, Zhang G, Liu D, Zhang J, Vang S, et al. 2006. High rate of chimeric gene origination by retroposition in plant genomes. *Plant Cell* **18**: 1791–1802. doi:10.1105/tpc.106.041905

Wetaya R. 2022a. Zambia's traditional pig farmers poised to embrace reproductive biotechnology in break from European-inspired anti-biotech sentiments. https://geneticliteracyproject.org/2022/09/06/zambias-traditional-pig-farmers-poised-to-embrace-reproductive-biotechnology-in-break-from-european-inspired-anti-biotech-sentiments

Wetaya R. 2022b. Rising above the high tide of anti-GMO misinformation in Africa: science experts and media join forces. https://allianceforscience.org/blog/2022/11/rising-above-the-high-tide-of-anti-gmo-misinformation-in-africa-science-experts-and-media-join-forces

Whelan A, Gutti P, Lema MA. 2020. Gene editing regulation and innovation economics. *Front Bioeng Biotechnol* **8**: 303. doi:10.3389/fbioe.2020.00303

Wood A. 2022. NC extension tips: ancient farmers who changed corn genes engaged in early GMO. https://www.dailyadvance.com/features/columnists/nc-extension-tips-ancient-farmers-who-changed-corn-genes-engaged-in-early-gmo/article_246fde4d-3709-5e4e-b081-dd49551e6566.html

World Trade Organization. 1994. Agreement on the application of sanitary and phytosanitary measures. https://www.wto.org/english/docs_e/legal_e/15sps_01_e.htm

Yarnall MTN, Ioannidi EI, Schmitt-Ulms C, Krajeski RN, Lim J, Villiger L, Zhou W, Jiang K, Garushyants SK, Roberts N, et al. 2023. Drag-and-drop genome insertion of large sequences without double strand DNA cleavage using CRISPR-directed integrases. *Nat Biotechnol* **41**: 500–512. doi:10.1038/s41587-022-01527-4

Yong E. 2014. How an antibiotic gene jumped all over the tree of life. https://www.nationalgeographic.com/science/article/how-an-antibiotic-gene-jumped-all-over-the-tree-of-life

Yong E. 2017. How a quarter of cow DNA came from reptiles. https://www.theatlantic.com/science/archive/2017/10/how-a-quarter-of-the-cow-genome-came-from-reptiles/542868

Zahoor S. 2022. Tinkering with saffron's genetic makeup to develop climate resilience. https://india.mongabay.com/2022/11/tinkering-with-saffrons-genetic-makeup-to-develop-climate-resilience

Zongo A. 2022. Agricultural biotechnologies: "Burkina Faso cannot be on the margins of progress in the world," Hamidou Traoré, Director of INERA. https://sentinellebf.com/biotechnologies-agricoles-le-burkina-faso-ne-peut-pas-etre-en-marge-du-progres-dans-le-monde-hamidou-traore-directeur-de-linera/

Cite this article as *Cold Spring Harb Perspect Biol* doi: 10.1101/cshperspect.a041677

Index